T0239039

On-Chip Photonic Interconnects: A Computer Architect's Perspective

Synthesis Lectures on Computer Architecture

Editor
Mark D. Hill, *University of Wisconsin*

Synthesis Lectures on Computer Architecture publishes 50- to 100-page publications on topics pertaining to the science and art of designing, analyzing, selecting and interconnecting hardware components to create computers that meet functional, performance and cost goals. The scope will largely follow the purview of premier computer architecture conferences, such as ISCA, HPCA, MICRO, and ASPLOS.

On-Chip Photonic Interconnects: A Computer Architect's Perspective
Christopher J. Nitta, Matthew K. Farrens, and Venkatesh Akella
2013

Optimization and Mathematical Modeling in Computer Architecture
Tony Nowatzki, Michael Ferris, Karthikeyan Sankaralingam, Cristian Estan, Nilay Vaish, and David Wood
2013

Security Basics for Computer Architects
Ruby B. Lee
2013

The Datacenter as a Computer: An Introduction to the Design of Warehouse-Scale Machines, Second edition
Luiz André Barroso, Jimmy Clidaras, and Urs Hölzle
2013

Shared-Memory Synchronization
Michael L. Scott
2013

Resilient Architecture Design for Voltage Variation
Vijay Janapa Reddi and Meeta Sharma Gupta
2013

Multithreading Architecture
Mario Nemirovsky and Dean M. Tullsen
2013

Performance Analysis and Tuning for General Purpose Graphics Processing Units (GPGPU)
Hyesoon Kim, Richard Vuduc, Sara Baghsorkhi, Jee Choi, and Wen-mei Hwu
2012

Automatic Parallelization: An Overview of Fundamental Compiler Techniques
Samuel P. Midkiff
2012

Phase Change Memory: From Devices to Systems
Moinuddin K. Qureshi, Sudhanva Gurumurthi, and Bipin Rajendran
2011

Multi-Core Cache Hierarchies
Rajeev Balasubramonian, Norman P. Jouppi, and Naveen Muralimanohar
2011

A Primer on Memory Consistency and Cache Coherence
Daniel J. Sorin, Mark D. Hill, and David A. Wood
2011

Dynamic Binary Modification: Tools, Techniques, and Applications
Kim Hazelwood
2011

Quantum Computing for Computer Architects, Second Edition
Tzvetan S. Metodi, Arvin I. Faruque, and Frederic T. Chong
2011

High Performance Datacenter Networks: Architectures, Algorithms, and Opportunities
Dennis Abts and John Kim
2011

Processor Microarchitecture: An Implementation Perspective
Antonio González, Fernando Latorre, and Grigorios Magklis
2010

Transactional Memory, 2nd edition
Tim Harris, James Larus, and Ravi Rajwar
2010

Computer Architecture Performance Evaluation Methods
Lieven Eeckhout
2010

Introduction to Reconfigurable Supercomputing
Marco Lanzagorta, Stephen Bique, and Robert Rosenberg
2009

On-Chip Networks
Natalie Enright Jerger and Li-Shiuan Peh
2009

The Memory System: You Can't Avoid It, You Can't Ignore It, You Can't Fake It
Bruce Jacob
2009

Fault Tolerant Computer Architecture
Daniel J. Sorin
2009

The Datacenter as a Computer: An Introduction to the Design of Warehouse-Scale
Machines
Luiz André Barroso and Urs Hölzle
2009

Computer Architecture Techniques for Power-Efficiency
Stefanos Kaxiras and Margaret Martonosi
2008

Chip Multiprocessor Architecture: Techniques to Improve Throughput and Latency
Kunle Olukotun, Lance Hammond, and James Laudon
2007

Transactional Memory
James R. Larus and Ravi Rajwar
2006

Quantum Computing for Computer Architects
Tzvetan S. Metodi and Frederic T. Chong
2006

© Springer Nature Switzerland AG 2022
Reprint of original edition © Morgan & Claypool 2014

All rights reserved. No part of this publication may be reproduced, stored in a retrieval system, or transmitted in any form or by any means—electronic, mechanical, photocopy, recording, or any other except for brief quotations in printed reviews, without the prior permission of the publisher.

On-Chip Photonic Interconnects: A Computer Architect's Perspective

Christopher J. Nitta, Matthew K. Farrens, and Venkatesh Akella

ISBN: 978-3-031-00646-3   paperback
ISBN: 978-3-031-01774-2   ebook

DOI 10.1007/978-3-031-01774-2

A Publication in the Springer series
SYNTHESIS LECTURES ON COMPUTER ARCHITECTURE

Lecture #27
Series Editor: Mark D. Hill, *University of Wisconsin*
Series ISSN
Synthesis Lectures on Computer Architecture
Print 1935-3235   Electronic 1935-3243

On-Chip Photonic Interconnects: A Computer Architect's Perspective

Christopher J. Nitta, Matthew K. Farrens, and Venkatesh Akella
University of California, Davis

SYNTHESIS LECTURES ON COMPUTER ARCHITECTURE #27

ABSTRACT

As the number of cores on a chip continues to climb, architects will need to address both bandwidth and power consumption issues related to the interconnection network. Electrical interconnects are not likely to scale well to a large number of processors for energy efficiency reasons, and the problem is compounded by the fact that there is a fixed total power budget for a die, dictated by the amount of heat that can be dissipated without special (and expensive) cooling and packaging techniques. Thus, there is a need to seek alternatives to electrical signaling for on-chip interconnection applications.

Photonics, which has a fundamentally different mechanism of signal propagation, offers the potential to not only overcome the drawbacks of electrical signaling, but also enable the architect to build energy efficient, scalable systems. The purpose of this book is to introduce computer architects to the possibilities and challenges of working with photons and designing on-chip photonic interconnection networks.

KEYWORDS

nanophotonics, on-chip network, interconnect, microring, optical interconnects, network topologies

Contents

List of Figures xiii

List of Tables xv

List of Acronyms xvii

Acknowledgments xix

1   Introduction 1
  1.1   Organization of the Lecture 5

2   Photonic Interconnect Basics 7
  2.1   Transmitter 7
       2.1.1 Lasers 7
       2.1.2 Microring Resonators 8
       2.1.3 Microrings As Modulators 12
  2.2   Transmission Medium 14
       2.2.1 Waveguide Details 14
       2.2.2 Vias 15
  2.3   Receiver 16
       2.3.1 Photodetector Details 17

3   Link Construction 19
  3.1   Photonic Link Design 19
       3.1.1 Transmitter 19
       3.1.2 Transmission Medium 20
       3.1.3 Receiver 20
       3.1.4 Design Decisions 21
       3.1.5 Photonic Power Requirements 21
       3.1.6 Electronic Power Requirements 24
       3.1.7 Layout/Implementation Issues 26
       3.1.8 Wide and Slow or Narrow and Fast? 28
       3.1.9 Total power 31

4   On-Chip Photonic Networks **35**
   4.1   Photonic Network Design Challenges 35
      4.1.1   Buffering 35
      4.1.2   Topology 35
      4.1.3   Arbitration and Flow Control 36
      4.1.4   Electrical/Optical Codesign 37
      4.1.5   Latency 37
   4.2   Case Studies of On-Chip Photonic Networks 38
      4.2.1   Corona 38
      4.2.2   Phastlane 41
      4.2.3   Firefly 44
      4.2.4   FlexiShare 45
      4.2.5   DCAF 46
      4.2.6   Hybrid Photonic NoC 48

5   Challenges **51**
   5.1   Process Variations 51
   5.2   Thermal Issues 52
   5.3   Trimming 53
   5.4   Resilient On-Chip Photonic Networks 54
      5.4.1   Photonic Link Fault Models 54
      5.4.2   Link Component Structure-Dependent Errors 56
      5.4.3   Unidirectional Bit Errors 58
      5.4.4   Mean Time Between Failures 59

6   Other Developments **63**
   6.1   On-chip Network Developments 63
      6.1.1   Monolithic CMOS Integration 63
      6.1.2   Lasers 64
      6.1.3   Plasmonics 65
   6.2   System-level Developments 67
      6.2.1   Off-chip I/O 67
      6.2.2   Memory System 69
      6.2.3   Large-Scale Routers 70

7   Summary & Conclusion **73**
   7.1   Observations and Things to Remember 73

Bibliography 77

Authors' Biographies 91

List of Figures

1.1 High-level photonic communication link 1

1.2 Latency optimized on-chip interconnect comparison 3

2.1 Laser spectrum 8

2.2 Microring micrograph 9

2.3 2x2 crossbar/microring comparison 9

2.4 Microring resonator and spectrum 10

2.5 Four cascaded microrings 11

2.6 Multiple microring spectrum 11

2.7 Passive and active microring resonators 12

2.8 Current injection spectrum 13

2.9 Waveguide configurations 14

2.10 Inter-layer couplers 16

3.1 Optical link example 20

3.2 Relative sizes of electrical and photonic components 25

3.3 Microring micrograph with buffer overlay 25

3.4 Alternating microring layout 26

3.5 Signal quality by bandwidth 27

3.6 Modulation and SERDES power 28

3.7 Microring wiring 29

3.8 Local transport power 30

3.9 Link power and energy efficiency (end) 30

3.10 Energy efficiency (fJ/b) vs. link utilization 32

3.11 Optical/electrical energy efficiency crossover 33

3.12 Electrical interconnect latencies vs. link length 34

4.1   MWSR bus and serpentine waveguide layout 39

4.2   Phastlane optical switch 42

4.3   SWMR buses 43

4.4   FlexiShare crossbar 45

4.5   DCAF TX 47

4.6   Hybrid NoC node, torus, and timing 49

5.1   Microring resonance vs. wavelength 53

5.2   Microring drop spectrum 55

5.3   Microring drop spectrum for interfering & non-interfering 55

5.4   Single-bit errors for faulty microrings 56

5.5   Double-bit errors for faulty microrings 57

5.6   Reed Solomon circuit 60

5.7   Microring fault rate for 1M hr MTBF 61

6.1   Hybrid plasmonic modulator 66

6.2   All-to-all interconnectivity in 5×5 AWGR 70

List of Tables

1.1   Evolution of photonic interconnects 2

3.1   Optical loss parameters from literature 22

3.2   Most energy efficient configuration 31

List of Acronyms

ACK	ACKnowledgement	42
ARQ	Automatic Repeat reQuest	42
AWGR	Arrayed Waveguide Grating Router	71
BOX	Buried Oxide	63
CMOS	Complementary Metal–Oxide–Semiconductor	4
CMP	Chip MultiProcessor	50
CRC	Cyclic Redundancy Check	59
DCAF	Directly Connected Arbitration-Free	46
DCOF	Directly Connected Optical Fabric	48
DIMM	Dual In-line Memory Module	69
DMA	Direct Memory Access	50
DWDM	Dense Wavelength Division Multiplexing	9
EDP	Energy Delay Product	45
FBDIMM	Fully Buffered DIMM	69
FEC	Forward Error Correction	59
FSR	Free Spectral Range	10
GBN	Go-Back-N	47
GF	Galois Field	61
HARQ	Hybrid Automatic Repeat reQuest	59
InP	Indium-Phosphide	64
ITRS	International Technology Roadmap for Semiconductor	2
LED	Light Emitting Diode	5
LFSR	Linear Feedback Shift Register	59
LVDS	Low Voltage Differential Signaling	60
MCM	Multi-Chip Module	67
MTBF	Mean Time Between Failure	61

MWSR	Multiple Writer Single Reader	39
MZI	Mach-Zehnder Interferometer	52
NAK	Negative AcKnowledge	36
NcK	N choose K	60
OOK	On-Off Keying	1
PIDRAM	Photonically Interconnected DRAM	70
PMMA	Polymethyl Methacrylate	52
PSE	Photonic Switching Element	48
QoS	Quality of Service	45
RAID	Redundant Array of Independent Disks	60
RSOA	Reflective Semiconductor Optical Amplifier	71
SAW	Stop-And-Wait	42
SECDED	Single Error Correction and Double Error Detection	60
SERDES	SERializer/DESerializer	20
SOI	Silicon-On-Insulator	2
SWMR	Single Writer Multiple Reader	44
TDP	Thermal Design Power	3
TED	Triple Error Detection	60
TIA	Transimpedance Amplifiers	16
TO	Thermo-Optic	52
TPA	Two-Photon Absorption	15
TSV	Through Silicon Via	4
UV	Ultraviolet	51
VCSEL	Vertical-Cavity Surface-Emitting Laser	5
VLSI	Very-Large-Scale Integration	2
WDM	Wavelength Division Multiplexing	4

Acknowledgments

The authors would like to express their gratitude to Christopher Batten of Cornell University for his detailed and thoughtful comments on the first draft of this book. We also thank Ben Yoo, Raj Amirtharajah, Linje Zhou, Amit Hadke, Paul Mejia, Stevan Djordjevic, and Roberto Proietti for the many lively discussions over the years that enhanced our understanding of some of the topics covered in this book. Support from the National Science Foundation through CCF Award 1116897 is gratefully acknowledged.

Christopher J. Nitta, Matthew K. Farrens, and Venkatesh Akella
September 2013

CHAPTER 1

Introduction

In this Synthesis Lecture we will provide an overview of photonic interconnects from the perspective of computer architects. Our objective is to enable someone who is familiar with electronic interconnects and computer architecture to quickly get an idea of the design space and landscape of photonic interconnects and become familiar with both the unique capabilities and challenges of using photonic interconnects in on-chip interconnection networks.

Photonics have been used in long distance communication for decades, so the underlying physics of the fundamental building blocks (lasers, waveguides, photonic switching, etc.) is well understood. Figure 1.1 shows a high-level illustration of a generic photonic link, as well as an example schematic enlargement which identifies the key components. In order to get the data (which is stored in electical form at the sender) to the receiver, the data must be modulated onto the light. There are many ways to modulate signals, with the simplest being On-Off Keying (OOK), where the presence/absence of light in a time interval encodes a 1 or a 0. A simple analogy is the telegraph—some text written on a piece of paper is handed to the keycode operator, who turns that hard copy text into a sequence of electrical signals (represented in the figure as data entering and exiting the transmitter). In the generic case the modulated light is coupled to a transmission medium, whose job is to transport the optical signal to the receiver. The receiver typically employs filters to separate individual communication channels and detectors to convert light back to an electrical signal.

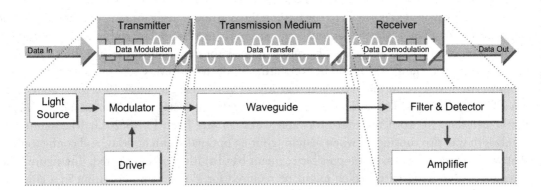

Figure 1.1: High-level photonic communication link with example photonic schematic enlargements.

Table 1.1: Evolution of photonic interconnects as a function of time and distance (data from Benner et al. [2005] and Hemenway [2013])

Interconnect	Length	Lines/Link	Lines/System	Time Frame
MAN/WAN	Multi-km	1	10s	Since 1980s
Cables-Long	10-300m	1 to 10s	10s to 1,000s	Since 1990s
Cables-Short	1-10m	1 to 10s	10s to 1,000s	Around 2005
Card-to-Card	0.3-1m	1 to 100s	10s to 1,000s	Around 2010
Intra-Card	0.1-0.3m	1 to 100s	1,000s	Around 2015
Intra-Module	5-100mm	1 to 100s	~10,000	Around 2020
Intra-Chip	0-20mm	1 to 100s	100,000s	Around 2025

Table 1.1 shows that the distances over which photonics make sense have continued to shrink, and over the years photonics have found their way into more and more areas of high-performance computing [Benner et al., 2005; Hemenway, 2013]. For example, photonics are now used to provide rack-to-rack, board-to-board and to some extent chip-to-chip connectivity, and may soon be used in datacenter networks [Farrington et al., 2010; Vahdat, 2012; Vahdat et al., 2011]. Supercomputers by Cray [Kim et al., 2008, 2009] and IBM [Arimilli et al., 2010] are already using dedicated photonic links in the interconnect hierarchy, and IBM researchers describe the feasibility of the underlying photonic building blocks and their role in exascale computing systems in Coteus et al. [2011]. At the chip-level, Intel researchers [Young et al., 2010] have demonstrated the advantages of using optical I/O to overcome the pin limitations of Very-Large-Scale Integration (VLSI) chips.

With technology scaling and Moore's Law likely to continue for the forseeable future, both the latency and power consumption of on-chip interconnects are becoming a major concern. The problem can be seen by looking at Figure 1.2, where we show the latency and associated energy consumption of both an electrical and photonic interconnect, plotted as a function of distance at the 16nm Silicon-On-Insulator (SOI) technology node using data from the International Technology Roadmap for Semiconductor (ITRS) [Semiconductor Industry Association, 2011]. The optimal electrical delay was calculated by finding the minimum delay for each number of possible segments (up to a point where additional segments only increase delay) and using the minimum of all the segment counts analyzed. The minimal delay for each segment length (link length divided by segment count) was found by comparing all delays for every repeater size between the minimal size inverter and the maximum size driver (which was capped at 100x the feature size). An RC model was used to calculate the segment switching delay, with the resistance and capacitance values for both the transistors and wires coming from ITRS 2011. Maximum wire bandwidth was also considered when searching for an optimal segment delay, and configurations that would exceed the maximum possible segment bandwidth were eliminated. The figure indicates that a 2cm-long wire (which might be required for global communication in a chip, for example) has a minimum latency of nearly 10 nanoseconds and requires approximately 1 picojoule per bit to transmit at that rate. In typical computing applications the interconnect data paths

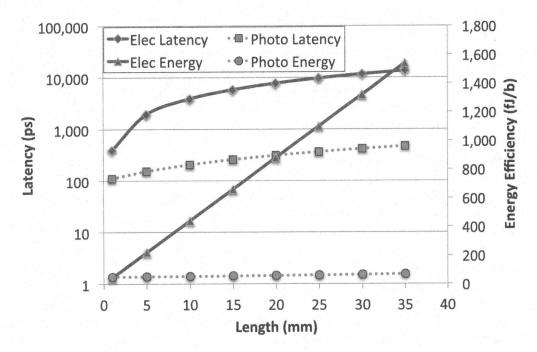

Figure 1.2: Latency optimized on-chip interconnect comparison at the 16nm technology node.

are 64–128 bits wide, so the energy consumed by the on-chip interconnect can quickly become a substantial component of the total energy budget of the chip. This leaves less energy for doing actual computations, because the total power that can be dissipated by a chip (called the Thermal Design Power (TDP)) is dictated by the packaging requirements and is a constant.

The potential latency and energy consumption of an on-chip photonic link is shown with dotted lines on Figure 1.2. The light source is assumed to be off-chip, but all of the other components from Figure 1.1 are on-chip, and the photonic link data was computed using the following parameters:[1] a 10fF photodetector (a receiverless design), 1V photodetector output, a 10fF microring, a 3.5V source driving the microring, 3dB Photodetector efficiency, 1dB/cm waveguide attenuation, 0.5dB on-resonance attenuation, and 0.0015dB off-resonance attenuation. The link was optimized for 10Gb/s, the rings were assumed to be athermal [Zhou et al., 2009b], and the propagation delay through the waveguide was assumed to be 10.45ps/mm. Figure 1.2 shows that the latency of the 2cm photonic interconnect is approximately 500ps, with an energy consumption of just 50 femtojoules per bit.

The energy consumption values in Figure 1.2 were calculated excluding the cost of any external off-chip components. Neither the energy required by the off-chip laser nor the energy

---

[1]Do not be concerned if the parameters listed do not make sense—they were included for the benefit of those who are already familiar with this topic. Each of the parameters will be discussed individually during the course of this book.

lost by the electrical power supply is taken into account in these calculations. The assumption is simply that an external light source exists, in the same way electrical designers assume a power supply to the chip exists, without taking into account all the losses of AC to DC transformation and DC to DC conversion that are inevitable. We are fully aware that lasers are currently more inefficient than typical voltage regulators, and that the design and efficiency of the laser is an important topic. We will revisit this topic in more detail later in this book, and also discuss possible alternatives to the use of an external laser that are on the horizon.

Figure 1.2 shows that on-chip photonic interconnects certainly have the potential to alleviate the latency and power consumption concerns in future large multicore processors—thus, it makes sense to explore the possibility of replacing some (or all) of the electrical interconnect on a chip with photonics, in order to simultaneously reduce the power consumed and meet the increasing bandwidth requirements of a larger number of cores per chip. So the question is, how practical are on-chip photonic interconnects? There are a variety of reasons to be encouraged.

For example, advances in 3D packaging technologies using Through Silicon Vias (TSVs) [Jeddeloh and Keeth, 2012; Motoyoshi, 2009; Woo et al., 2010] will allow a separate die with photonic components to be tightly coupled to a conventional Complementary Metal–Oxide–Semiconductor (CMOS) microprocessor. This approach will provide significant cost advantages, as it will simplify the manufacturing and testing of chips that use photonic interconnects. In addition, the basic building blocks of photonic interconnection networks, such as switches and waveguides, can now be built with an area budget that is acceptable for on-chip applications. (Components used in off-chip photonic interconnects such as lasers, modulators, couplers, etc. are very large, and thus not directly suitable for an on-chip interconnection application.)

Figure 1.2 also shows that the energy efficiency and latency of photonic interconnects is relatively independent of distance. This is a huge advantage when compared to electrical signaling, where latency and energy efficiency deteriorate signicantly as distances increase. There are two other important factors to consider, which are not apparent in the figure: first, increasing the data rate in a photonic link does not result in a significant decrease in energy efficiency, since the photonic losses (which are the dominant component of the energy) are largely independent of data rate. (This is referred to as *bit-rate transparency* in the photonics community.) Second, photonics has a much higher bandwidth-density, or bandwidth per unit area. When using photonics, the same physical waveguide (the photonic counterpart of an electrical wire) can be used to transport multiple bits simultaneously, using a different wavelength of light for each bit (a technique known as Wavelength Division Multiplexing (WDM)). For example, if each wavelength operates at 20Gb/s and there are 16 wavelengths per waveguide, one can effectively get a 16-bit bus with a bandwidth of 320Gb/s using the same area as a single waveguide.

These unique advantages of photonic interconnects over electrical interconnects (better energy efficiency, lower link latency, bit rate transparency, and superior raw bandwidth and bandwidth density) stem from the differences in the physics of propogation of electrical signals vs. photons [Miller, 2000, 2009]. The high carrier frequency of optical signals allows them to be guided by extremely low-loss dielectric waveguides, instead of the metal waveguides (wires) that are required for electrical signals; hence, photonic signals do not suffer from propagation loss and distortion due to the resistance of the metal waveguides. They are also relatively immune to reflection and crosstalk (between waveguides). These factors combine to give photonic links the ability to communicate over longer physical distances and at significantly higher data rates without major latency or power penalties.

Furthermore, these unique characteristics of photonic interconnects can be harnessed to create computing systems that are not only highly scalable, but also perhaps easier to program. Since electrical interconnects do not scale with distance, hierarchical network topologies must be used, which force the programmer to be aware of the on-chip/off-chip dichotomy. This is especially important in the context of multicore processors, because some inter-processor and processor-memory communication is on-chip and some off-chip, and they have different communication latencies. But given its ability to efficiently communicate over longer distances, photonics allows the designer to use *flatter* networks, and/or networks with uniform communication latency (such as crossbars or fully connected topologies) without compromising scalability. In addition, when using photonic interconnects the on-chip/off-chip dichotomy is less important because the communication latency is not significantly different between the two.

1.1   ORGANIZATION OF THE LECTURE

As stated previously, Figure 1.1 shows a high-level illustration of a generic photonic link. For on-chip photonic networks, the required light source could be a laser, a Light Emitting Diode (LED), or a Vertical-Cavity Surface-Emitting Laser (VCSEL), and this light source can be either on or off-chip—in this book we will assume it is an external (off-chip) laser, because the technology already exists to realize this approach.[2] The transmission medium is a waveguide, which can be viewed as an on-chip photonic wire whose job is to transport the optical signal to the receiver. It is likely the receiver may also need to use amplifiers to boost electrical signals produced by the detectors that are too weak to drive standard gates. Though not explicitly shown in the figure, a clock signal is also necessary to ensure that the transmitter and the recever are synchronized.

In Chapter 2 we will describe each of the link components in detail, and outline both different ways to realize them efficiently and the key design parameters that affect their performance. We will follow that with Chapter 3, in which we will describe the design and optimization of a single photonic link, and identify the tuneable parameters (or "knobs") that influence the performance and power consumption of a link.

Once we more fully understand the design of a photonic link, we will address the question of how one builds a network using them. It is here that we will encounter a key challenge to using photonic interconnects—the fact that light cannot easily be buffered. Because of this fact, the widely understood store-and-forward paradigm used by electronic on-chip networks is

---

[2]In both Chapter 2 and in Chapter 6 we will discuss the difficulties of constructing on-chip lasers and describe recent advances in the area.

not directly applicable to on-chip photonic networks. It is not possible to take certain electronic on-chip network topologies and simply swap photonic links for electrical wires and expect to immediately reap the benefits of higher bandwidth and lower energy per bit offered by photonics.

In addition, as Chapter 3 will point out, issues such as link utilization and the length of a photonic link matter, and there are limits to the bandwidth that photonic links can deliver in an energy efficient manner. Thus, on-chip photonic interconnect network design has to be approached as a codesign of both electronics and photonics, and issues such as topology, arbitration, and flow control must be carefully evaluated. This is the subject of Chapter 4. We conclude Chapter 4 with examples of published photonic on-chip networks to illustrate how various researchers have dealt with these trade-offs.

In Chapter 5 we identify some of the key technical challenges that need to be overcome in order for photonic interconnects to be viable in practice. For example, photonic devices are very sensitive to variations in temperature, making them susceptible to malfunction. This is particularly important if a computing system design expects hundreds of thousands of these components to function correctly at all times. This is a huge challenge and we present a brief overview of some of the approaches proposed by the research community to address this problem. It is also true that the fabrication of photonic components at nanoscale geometries is a new field, so techniques for fault simulation, fault modeling, and testing need to be considered. We present fault models and metrics that can be used when creating resilient photonic links.

Though the focus of this book is on on-chip photonic interconnects, there are some exciting developments in related areas that should be of interest to a computer architect. In Chapter 6 we present a brief of overview of these developments. We conclude in Chapter 7 with a recap of the key assumptions, observations, and things that one should remember when contemplating the use of photonic interconnects in computer applications.

IN DEPTH

In this book we will include sections that contain more in-depth material, which is not necessary in order to gain a basic understanding of the subject at hand. We will mark these sections as being "In Depth"—these can be thought of as essentially large footnotes, and are targeted at those people who may need more detailed information (in order to write a simulator, for example.) The reader is free to skip over these with no loss of general understanding.

CHAPTER 2

Photonic Interconnect Basics

In this chapter we will cover the basics of the on-chip photonic interconnect. The specific implementations of photonic interconnects can vary substantially, but from a high-level view they all have three basic components: a transmitter, a transmission medium, and a receiver (as shown in Figure 1.1). The transmitter converts electrical data into lightwaves, which are carried by the transmission medium to the receiver, where they are converted back to electrical data. This process is very similar to what happens when data is sent via a modem over a normal phone line—however, instead of being converted to tonally modulated electrical signals, the data is converted to the presence or absence of photons.

2.1 TRANSMITTER

All photonic transmitters send data by modulating light, either directly or indirectly. Direct modulation of light occurs when the supply of power to the light source is modulated—turning a light switch on and off to send a signal, for example. Indirect modulation of light is when the light source is constant and the light is externally modulated—think of a signal lamp used for ship to ship communication. The light source is constantly on, and the light is shuttered (modulated) to send information. In this work we will focus on indirect modulation using microring resonators and an external laser source for two reasons—microring resonators are area-efficient and thus suitable for on-chip photonic interconnects, and external lasers are already commercially available.

2.1.1 LASERS

In order to understand the rational for using external lasers[1] we must first look at the materials used for semiconductors. Semiconductor materials can be classified as having either a direct or an indirect band gap. In direct band gap materials, electrons can directly emit a photon because the momentum of the electrons and holes is the same for both the conduction and valence bands. In indirect band gap materials, electrons must first transfer their momentum to an intermediate state before emitting a photon – this makes indirect band gap materials much less efficient at providing a stimulated emission of photons (a less ideal choice for a light source, in other words). Since silicon is an indirect band gap material, the majority of on-chip nanophotonic research has focused on the use of an external light source to act as an "external photonic battery" [Lipson, 2005]; furthermore, researchers have primarily assumed the use of "off the shelf" lasers which are

---

[1]As a reminder, "Laser" stands for Light Amplification by Stimulated Emission of Radiation

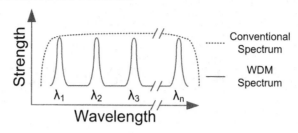

Figure 2.1: Conventional and WDM laser spectrum.

readily available and being used in telecommunications, so they could concentrate on the novel aspects of on-chip photonics.[2]

The most commonly used lasers in telecommunications are those in the C (or "conventional") band, which ranges from 1530 to 1565nm. These lasers attempt to provide a wide, flat-power spectrum, as is shown in Figure 2.1. While these readily available C band lasers are compatible with silicon nanophotonic structures, they do not produce the multiple discrete wavelengths needed for WDM. As we mentioned in the previous chapter, by using different wavelengths WDM allows multiple data bits to be sent simultaneously down a single transmission medium. A WDM spectrum is also shown in Figure 2.1. It is possible to couple multiple, narrow-band lasers together in order to provide the necessary set of wavelengths; however, multiple discrete wavelengths can also be produced using a single mode-locked laser [Koch et al., 2007], or by using a tunable laser with a comb filter [Sun et al., 2008]. While lasers that produce a WDM spectrum are commercially available, challenges such as reducing cost and power consumption (the thermal controls required to stabilize wavelength and photonic power output consume significant power) still remain [Liu et al., 2010a].

2.1.2 MICRORING RESONATORS

Microring resonators are WDM-compatible devices that are compact and energy efficient [Miller, 2009], and are designed to resonate when presented with specific individual wavelengths and remain quiescent at all other times. A micrograph of a 10μm microring resonator is shown in Figure 2.2. From a computer architect's perspective, a microring resonator is similar to a 2x2 crossbar switch in which only the straight and exchange states exist (we will discuss the mechanisms of switching the microring between these two states in Section 2.1.3). Figures 2.3(a) and (b) show a 2x2 crossbar and a microring resonator in the "straight" (or "off-resonance") state, with a logical one (the presence of wavelength λ) at the *input* port and a logical zero (the absence of wavelength λ) at the *add* port. Figures 2.3(c) and (d) show the "exchange" (or "on-resonance")

---

[2]Creating a monolithically integrated laser on silicon (i.e., an on-chip laser) is an active area of research, because it could be one of the enabling technologies that enables/accelerates the use of photonics on-chip. There are reasons to believe that on-chip lasers may one day be a reality; please see Chapter 6 for further details.

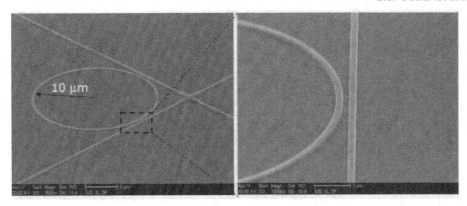

Figure 2.2: Micrograph of an actual 10 m diameter microring. This is a passive ring and would be used to move certain wavelengths from the waveguide running at a 45 angle to/from the one perpendicular to it. Note that the waveguide is approximately 400nm from the microring itself.

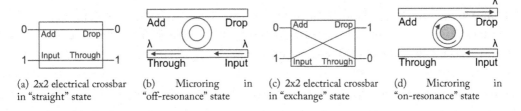

(a) 2x2 electrical crossbar in "straight" state

(b) Microring in "off-resonance" state

(c) 2x2 electrical crossbar in "exchange" state

(d) Microring in "on-resonance" state

Figure 2.3: 2x2 electrical crossbars in "straight" (a) and "exchange" (c) states, and microring resonantor in "off-resonance" (b) and "on-resonance" (d) states.

states. Every individual microring is capable of modulating a single bit (a single wavelength); thus, a transmitter with a multi-bit, parallel data path can be constructed using multiple microrings and a WDM-capable light source.

Microring Resonator Details
The particular wavelength that a microring resonates to is dependent upon both the microring radius and its effective index of refraction (see the "In Depth: Index of Refraction" section regarding the index of refraction for further details). The ability to respond to specific wavelengths enables the removal (filtering) of specific wavelengths from a waveguide, and these resonators are the primary technology used to bundle the high quantity of wavelengths per waveguide needed when using Dense Wavelength Division Multiplexing (DWDM). Figures 2.4(a) and 2.4(b) show microrings that resonate at $_1$ being used in perpendicular and parallel waveguide configurations (respectively). The specific wavelength that a microring is tuned to resonate to moves the wave-

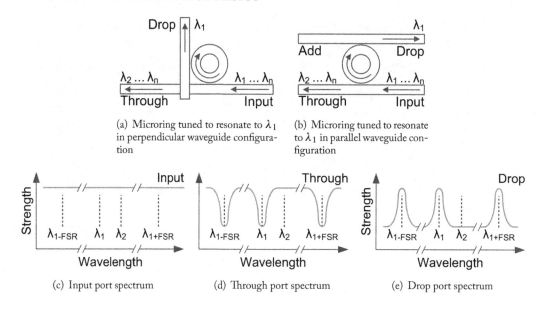

(a) Microring tuned to resonate to λ_1 in perpendicular waveguide configuration

(b) Microring tuned to resonate to λ_1 in parallel waveguide configuration

(c) Input port spectrum

(d) Through port spectrum

(e) Drop port spectrum

Figure 2.4: Microring resonators: (a) and (b) show microrings, which at fabrication time were set to resonate only to λ_1. The microring resonator spectrum at the through (d) and drop (e) ports is shown when a constant power spectrum is applied at the input (c) port.

length from the *input/through* waveguide to the *add/drop* waveguide. This wavelength transistioning mechanism is symmetric—for example, if λ_1 entered the *add* waveguide in Figure 2.4(b), it would transition to the other waveguide and exit at *through*. For the sake of understanding, computer architects may view this wavelength transitioning mechanism as discrete (i.e., only λ_i will transition and no other wavelengths will)—however, microring resonators are in reality simultaneously both continuous band pass and band reject filters. Figures 2.4(d) and 2.4(e) illustrate the spectrum for the *through* (band reject) and *drop* (band pass) respectively, provided that a constant power spectrum is applied at the *input* (illustrated in Figure 2.4(c)). The reader may notice that there are multiple peaks or troughs in Figures 2.4(d) and 2.4(e); the resonance wavelengths to which microring resonators respond repeat at an interval known as the Free Spectral Range (FSR)—for example, a microring designed to resonate at λ_1 will also resonate at $\lambda_1 \pm$ FSR, $\lambda_1 \pm 2$FSR, etc.

IN DEPTH: MICRORING RESONATORS

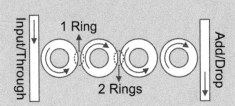

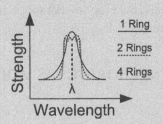

Figure 2.5: Four cascaded microrings.   **Figure 2.6:** Multiple microring spectrum.

The pass band created by the microring matches a near-Lorentzian function, and this function, combined with the FSR of the microrings, imposes a limit on the number of wavelengths that can be used in DWDM (primarily due to crosstalk issues [Sherwood-Droz et al., 2010]). Crosstalk is any phenomenon by which a signal transmitted on one circuit or channel of a transmission system creates an undesired effect in another circuit or channel, and in photonics it is related to the width and sharpness of the pass band.

It is possible to create filters with sharper roll-off and a flatter top by using cascading microrings. Figure 2.5 shows four cascaded microrings and Figure 2.6 presents the filter spectrum for 1, 2, and 4 cascaded microrings. Notice that as the number of microrings increases the roll-off becomes sharper and a "flatter" top with multiple peaks occurs, but at a cost of peak resonance strength. Depending upon the interconnect design, this additional signal attenuation may be a reasonable sacrifice in order to provide reduced signal crosstalk.

The 3dB bandwidth of the microring filters defines the limit of the wavelength communication channel. The Shannon-Hartley theorem states that $C = B log_2(1 + \frac{S}{N})$, where C is the channel capacity (in bits per second), B is the channel bandwidth, and $\frac{S}{N}$ is the signal-to-noise ratio. Given the 3dB bandwidth B_{3dB}, then $C = B_{3dB} log_2(1 + 10^{\frac{3}{10}}) \approx 1.5 B_{3dB}$, which means that the 3dB bandwidth is roughly equal to the channel capacity—thus, creating a sharper roll-off can create a wavelength filter with a higher, wider 3dB pass band and hence a larger channel capacity. The 3dB bandwidth of a microring is often represented in the form of the microrings' Q factor. The Q factor is the quality factor for a resonator, and is written $\frac{f_0}{\Delta f}$ (where f_0 is the center frequency and Δf is the 3dB bandwidth of the resonator). Assuming a 1550nm laser, the f_0 would be 193.5THz ($\frac{c}{1550nm}$) and if Δf of ~20GHz is desired, the Q factor would be approximately 10,000. The Q target of the microring also dictates the limit of the diameter of a ring, according to Xu et al. [2008]; "For a modulator working at 10-20 Gbit/s, a moderately high operating Q on the order of 10,000, which corresponds to an optical bandwidth of ~20 GHz, is appropriate for the critically coupled resonator, which requires an intrinsic Q of 20,000. ...one can conclude that the minimal radius to obtain an intrinsic Q of 20,000 around the wavelength of

1.55nm is 1.37um." In other words, for a 193.5THz f_0 (1550nm laser), a microring diameter of 2.74um is the minimum that can be used while maintaining an acceptable Q factor. The Q factor is exponentially proportional to the microring radius, while the FSR is inversely proportional to the microring radius.

2.1.3   MICRORINGS AS MODULATORS

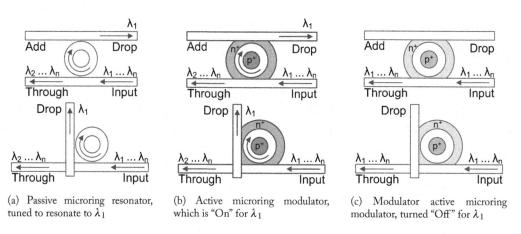

(a) Passive microring resonator, tuned to resonate to λ_1

(b) Active microring modulator, which is "On" for λ_1

(c) Modulator active microring modulator, turned "Off" for λ_1

Figure 2.7: Microring resonators: (a) shows a passive microring, which at fabrication time was set to resonate only to λ_1. (b) and (c) show active microrings, which use the presence or absence of charge in the n^+ base to change the wavelength to which they will resonate to (λ_1, here).

The wavelength filtering described previously can be achieved using either passive or active microrings. Figure 2.7(a) shows a passive microring that is biased during fabrication to extract only λ_1 from one waveguide and steer it down the other. Since the passive microrings are biased during fabrication to always respond to a single wavelength, they cannot be used for modulation. Modulation requires an active microring resonator, which is designed to change its resonance frequency via either carrier injection (in a silicon PIN diode) or carrier depletion (in a silicon PN diode[3]). In other words, an active resonator uses the presence or absence of charge in the n^+ region to change the wavelength to which they will resonate. Though carrier depletion requires lower voltages and can lead to faster devices [Feng et al., 2010], carrier injection can induce a larger change in the index of refraction, allowing the resonance wavelength to be shifted further. This ability can potentially simplify receiver design, since it enables the difference between the on and off power levels to be greater, although this advantage comes at a cost of slower and possibly less efficient modulating devices. Clearly there are tradeoffs to using each approach—however,

---

[3]Modulators have also been constructed using an MOS-capacitor combined with the electro-absorption effect.

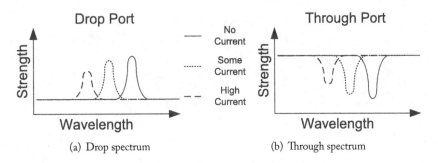

Figure 2.8: Spectrum of drop (a) and through (b) for a microring under varying amounts of current injection.

because of the potential to simplify the design of the receiver, in the remainder of this book we will assume carrier injection (also referred to as current injection) is used for modulation.

Figure 2.7(b) illustrates an active microring resonator modulating wavelength λ_1. In this figure it is assumed that if an electrical current is present in the n^+ base, λ_1 is extracted and sent down the *drop* waveguide—if there is no current applied (as in Figure 2.7(c)), λ_1 will continue down the *input/through* waveguide unaffected. (Though the figure shows a microring that is "on" when current is applied, the microring can be designed to do the opposite—it can be "on" resonance when current is not applied and be "off" resonance when current is injected.)

Injecting current into the microring base causes signal attenuation when transitioning wavelengths, something the architect/designer must keep in mind. Figures 2.8(a) and 2.8(b) show the spectrums for the drop and through ports under various levels of current injection, assuming a continuous spectrum of wavelengths is applied at the input port. Notice that as more current is injected, the resonance both shifts left toward the blue (higher frequency/shorter wavelength) and the amplitude decreases (the signal deteriorates). The architect/designer must also deal with the fact that microrings are thermally sensitive—as they heat up, their resonance shifts toward the red. Generally speaking, dealing with this thermal sensitivity is addressed using a technique called "trimming", which we will discuss in detail in Chapter 5.

The method by which an active microring modulates depends upon the configuration of the incoming and outgoing waveguides. If it is assumed that the presence of a wavelength represents a logic 1 and the absence represents a logic 0 (OOK), and if the incoming waveguide is also the outgoing waveguide, then a zero can be created by using the microring to remove the wavelength by bending it onto a dead-end drop waveguide (as shown in Figure 2.7(b)) and a one created by allowing the wavelength to pass unaffected (as is shown in Figure 2.7(c)). In the remainder of this book we will refer to this approach as modulating zeros, since the wavelengths must actively be removed in order to create a zero. If the incoming and outgoing waveguides are not the same, then ones are created by bending the wavelength onto the outgoing waveguide (as in Figure 2.7(b)), and zeros by allowing the wavelength to continue unperturbed along the incoming waveguide (shown

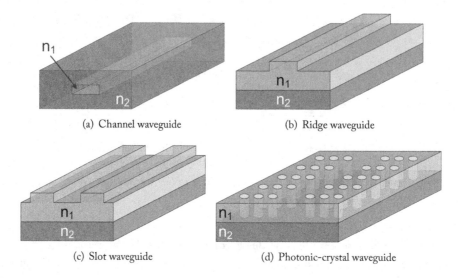

Figure 2.9: Channel (a), ridge (b), slot (c), and photonic-crystal (d) waveguide configurations.

in Figure 2.7(c)). As one might expect, we will refer to this approach as modulating ones, since the wavelengths must actively be transitioned in order to create a one. There are definite advantages and disadvantages to either approach—for example, when modulating ones, the photonic path will incur additional attenuation due to wavelength transitioning. However, we will show in Section 5.4 that there are significant advantages with respect to system resilience if modulating ones is employed. There have been designs presented in the literature that used one or the other of these approaches, and one (Corona [Vantrease et al., 2008]) that actually used both methods in a single network—modulation of zeros was used for the data path, and modulation of ones was used for the arbitration system.

2.2 TRANSMISSION MEDIUM

There are a number of different media which can be used to carry photons between the transmitter and receiver in an on-chip photonic interconnect. In this book we will assume the transmission medium is part of an on-chip waveguide, and is most likely silicon (although the waveguide could be air, SiO_2, Si_3N_4, etc).

2.2.1 WAVEGUIDE DETAILS

There are four configurations of waveguides in silicon: channel, ridge, slot, and photonic-crystal. These four are shown in Figure 2.9. Channel and ridge waveguides (Figures 2.9(a) and 2.9(b), respectively) are the most common, and rely on total internal reflection which concentrates the light in the high index of refraction material. These waveguide configurations provide single-mode propagation, which is desirable since multi-mode propagation can suffer from modal dispersion.[4] Slot and photonic-crystal waveguides (Figures 2.9(c) and 2.9(d), respectively) are less common configurations and confine light using mechanisms other than total internal reflection.

IN DEPTH: INDEX OF REFRACTION

The index of refraction of a material (or optical medium) n is equal to $\frac{c}{v}$, where c is the speed of light in a vacuum and v is the speed of light in the medium. Total internal reflection occurs when the incident angle of light (measured from the surface normal) is greater than the critical angle $\theta_c = \arcsin \frac{n_2}{n_1}$, where n_1 is the index of refraction of the optical medium, and n_2 is the index of refraction of the cladding material. The index of refraction for materials is dependent upon the wavelength of light λ, and can be determined using the Sellmeier equation $n^2(\lambda) = 1 + \frac{B_1\lambda^2}{\lambda^2 - C_1} + \frac{B_2\lambda^2}{\lambda^2 - C_2} + \frac{B_3\lambda^2}{\lambda^2 - C_3}$, where coefficients $B_{1,2,3}$ and $C_{1,2,3}$ are experimentally determined for each material.

There are four sources of propagation loss in on-chip waveguides: the intrinsic absorption of the material, radiative coupling, scattering, and Two-Photon Absorption (TPA). The intrinsic absorption of light in Si and SiO_2 at the wavelengths used for nanophotonics is negligible when compared to other sources of loss [Lipson, 2005], while radiative coupling losses occur due to rapid changes in the waveguide geometry (as occurs in tight bends). Scattering loss due to roughness of the waveguide sidewall is the main source of propagation loss in on-chip waveguides, while the least common source of loss is due to TPA: TPA occurs in silicon waveguides when excessive photonic power is fed into the waveguide and can contribute to signal crosstalk in DWDM [Sherwood-Droz et al., 2010; Tsang et al., 2002].

Waveguides can intersect without complete signal interference, unlike wires carrying electronic signals. When the angle of intersection is 90 degrees, the interference is minimized—signals traveling down each waveguide will continue on intact, although each signal will suffer a small attenuation (often assumed to be between 0.05dB [Joshi et al., 2009a; Pan et al., 2010] and 0.2dB [Koka et al., 2012]). There may also be signal crosstalk and partial signal reflection. This characteristic of photonics has allowed on-chip optical networks to be laid out on a single layer—however, the cumulative effect of a large number of intersections may make a single layer waveguide layout infeasible. Waveguides may need to be routed on different layers in order to avoid excessive intersections.

2.2.2   VIAS

In the electronic domain signals can easily move from layer to layer using vias, and transitioning photonic signals from one layer to another is done in a similar manner. Grating couplers are used to couple optical fibers and waveguides [Maire et al., 2008; Taillaert et al., 2004], and vertical grating couplers can be used to connect waveguides on different layers. Figure 2.10(a)

---

[4]Modal dispersion occurs because the propogation velocity of different modes is not the same.

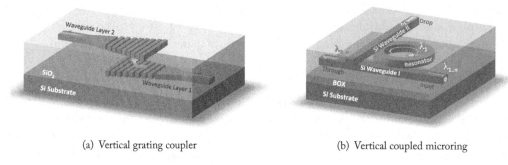

(a) Vertical grating coupler   (b) Vertical coupled microring

Figure 2.10: Inter-layer couplers (photonic vias): a vertical grating coupler (a) and a vertically coupled microring (b).

illustrates a grating coupler being used as a photonic via—this transitions all wavelengths from the waveguide on one layer to the waveguide on another. Microring resonators are also capable of vertical coupling, though they can only transition a single wavelength. The microring shown in Figure 2.10(b) is vertically coupled, allowing perpendicular waveguides on seperate layers to be coupled by a single microring, eliminating the need to actually intersect.

2.3   RECEIVER

The receivers used in on-chip photonic interconnects are constructed using microrings, photodetectors, and an amplifier. The microrings are used to extract the individual wavelengths from the waveguide and direct the extracted wavelengths to photodetectors, which convert the photonic power into an electrical signal. In many cases the voltage output from the photodetector requires amplification, which is typically provided by a Transimpedance Amplifiers (TIA) (although simpler designs, such as resistive or current integrating receivers, have been proposed [Georgas et al., 2011]). Designs which include an amplification stage must be concerned with the energy consumed each time the amplifier signal is switched, as well as the static power consumed by the amplifier. Amplification may not be neccessary if the capacitance of the photodetectors can be reduced to a low enough level (on the order of a few femtofarads); if this is possible, a "receiverless" approach can be used [Bhatnagar et al., 2002; Miller, 2009]. In our analysis presented in later chapters of this book, we assume that TIAs (or similar amplification stages) are not needed and that a "receiverless" approach is possible. We make this assumption for two reasons: first, not having to include the optimal design of an amplifier greatly reduces an already extremely large design space exploration, and second, while it is somewhat optimistic to assume receiverless designs, it is less so than other proposals presented in the literature (for example, optical nano-antennas [Cao et al., 2010; Yousefi and Foster, 2012] which may not even require photodetectors.). Receiverless designs are also assumed in many published works [Ahn et al., 2009; Nitta, 2011; Nitta et al., 2012c; Vantrease, 2010].

2.3.1 PHOTODETECTOR DETAILS

A photodiode is a quantum detector that converts optical energy into electrical energy, and the amount of incident optical energy required to generate the desired output voltage is critical to the energy efficiency of the entire photonic link. This is due to the fact that the energy required at the light source is $10^{\frac{A}{10}}$ times that of the energy required at the photodetector (where A is the attenuation of the path measured in decibels). It is not uncommon for researchers to propose networks that have path attenuations in the range of 10 to 20dB, meaning that the required energy at the light source will be 10 to 100 times that which reaches the final photodetector. PIN photodiodes/phototransistors are most often proposed for use as photodetectors in on-chip photonic interconnects [Kirman et al., 2006].

IN DEPTH: PHOTODETECTOR EFFICIENCY

The optical energy E_L required to generate a photodetector output voltage of V_0 (in the static case) is given by the equation $E_L = \frac{\hbar\omega V_0 C_D}{\beta e}$, where $\hbar$ is the reduced Planck constant, ω is the photon angular frequency, C_D is the capacitance of the photodetector, β is the photodetector quantum efficiency, and e is the electronic charge [Miller, 1989]. This can also be written $E_L = \frac{hc V_0 C_D}{\lambda\beta e}$, where h is the Planck constant, c is the speed of light, and λ is the photon wavelength. This holds for the dynamic case as well, assuming that $RC_D \geq \tau$, where R is the diode-loaded resistance and τ is the light-pulse duration. It should be clear that the required optical energy is proportional to both V_0 and C_D—thus, reducing the capacitance of the photodetector will directly improve the energy efficiency of the photonic link. If it is reduced to a sufficiently low level, a "receiverless" approach can be used [Miller, 2009].

In the case where C_D cannot be reduced, V_0 can be reduced and an amplifier stage can be used to generate the voltages neccessary to drive a gate. A TIA, resistive receiver, current integrating receiver [Georgas et al., 2011], or some other amplification stage will be necessary to convert the photodetector current to a voltage capable of switching subsequent digital gates. A TIA provides a gain $K = \frac{V_{out}}{I_{in}}$ measured in Ωs, which is typically desired to be as large as possible, though there are many other competing design constraints (such as providing necessary 3dB bandwidth, reducing amplifier power consumption, reducing noise, and reducing group delay/latency [Pappu and Apsel, 2005]). Designing an optimal TIA is difficult because TIAs generally have complex transfer functions with multiple poles [Bespalko, 2007].

C H A P T E R   3

Link Construction

In Chapter 2 we introduced the basic components used in on-chip photonic interconnects. In this chapter we will present an example of how a link can be constructed using these components, and then look in detail at the parameters that influence the performance and power consumption of a photonic link. We will also describe the trade-offs between power consumption, the number of parallel bits used, and the signaling rate used to attain a target link bandwidth.

Some of these tradeoffs are different from those in traditional electrical link design, since Moore's Law does not apply to photonics in the same way it does to conventional CMOS-based electronics. Semiconductor microrings can be made with a diameter as small as 3μm, but that is close to the theoretical limit for a silicon microring resonator with an acceptable Q factor (see the "In Depth: Microring Resonators" in Chapter 2 for more details). Since the microrings are limited in how small they can get, but the supporting electronics are not, an imbalance is created. This is important because electrical signals are used to move data from the storage buffers to the photonic transmitters, to move incoming data from the photonic receivers to the receive buffers, and to provide the modulating signals to the microrings. When the electrical feature sizes are shrinking but the photonic rings are not, it becomes more and more challenging to get the data and control signals to and from the microrings in an energy-efficient manner. Thus, we will show that it is critical to take both the electronics and the photonics into account simultaneously, from the beginning, when designing an on-chip photonic network.

3.1   PHOTONIC LINK DESIGN

In this section we will present an example of a photonic link (shown in Figure 3.1) that directly connects source and destination nodes.

3.1.1   TRANSMITTER

As stated previously, we assume the link uses an external laser as a light source, which provides the necessary set of wavelengths used for communication. These wavelengths are delivered to the chip via an optical fiber, and once they are on chip a waveguide carries the wavelengths to the microring modulators. The transmitter consists of the combination of the external laser, optical fiber, waveguides, serializer, electrical drivers and microring modulators. The transmitter electrical interface is m bits wide, a number which is equal to that of the interconnect flit width; this m-bit

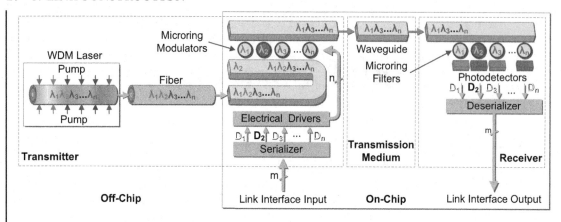

Figure 3.1: Example of an optical link with an m-bit flit and an n-bit phit, which is sending a phit data value 1...1101. White D_i labels denote a logical one while black D_i labels denote a logical zero.

flit is connected to an $m : n$ serializer, where n is the phit width.[1] The output of the serializer feeds into the electrical drivers, which convert the signals into drive currents that are injected into the microring modulators. This transitions the appropriate wavelengths to the outgoing waveguide (it is assumed that ones are actively modulated in this link)—in this case λ_1 and λ_3 through λ_n are transfered, creating the desired pattern for the phit.

3.1.2   TRANSMISSION MEDIUM

The wavelengths travel down the waveguide, which is the transmission medium from the source to the destination node. Figure 3.1 shows only a single waveguide between the transmitter and the receiver, although multiple waveguides could be used. We assume that the transmission medium section consists only of the waveguide, and that intermediate microrings are not used for routing.

3.1.3   RECEIVER

Once the transmitted value arrives at the receiver of the destination node, microrings remove the individual wavelengths from the incoming waveguide and route them to photodetectors, where the photonic power is converted back to electrical signals. As stated in the previous chapter, we assume that the photodetectors are of low enough capacitance that the link can be run "receiver-less," meaning that TIAs are not needed to amplify the signals from the photodetectors in order to drive the inputs of the $n : m$ deserializer (assuming SERializer/DESerializer (SERDES) is needed). If amplification *is* needed, there would be an additional TIA stage (or similar sense amplification) between the photodetectors and the deserializer. The phit transmission is complete

---

[1]A *flit* is a flow control digit, i.e., the smallest unit of flow control. A *phit* is the basic unit of data transfer at the physical layer; in other words, the number of bits used in each physical data transfer.

once the signals reach the deserializer, and the flit transmission is complete at the output of the deserializer at the receiver interface.

3.1.4   DESIGN DECISIONS

When designing an interconnect, the computer architect must decide the amount of bandwidth to provide, the energy cost of providing that bandwidth, and the amount of chip area to devote to the link. Since the area required for the link transmitter and receiver is dominated by the microring count, we will focus on optimizing the energy efficiency of the link. Bandwidth is the amount of information transferred per unit time,[2] and is a function of both the number of bits being sent during each transmission and the number of transmissions done per second. Choosing the number of parallel bits to use (the bits per transaction, or the phit width) and the signaling rate (the transactions per second) used to attain a target link bandwidth involves trade-offs in both power and area, since there is obviously an inverse relationship between phit width and signaling rate—as the phit width goes down, the signaling rate must go up.

In the next two subsections we will examine in more detail the photonic and electrical contributions to the total power used and how they are affected by the choice of phit width and frequency. We will provide experimental results for a 35mm-long link with a target bandwidth of 640Gb/s for 32, 22, and 16nm technology points. This configuration was chosen because it represents a 64-bit wide optical data path operated at 10GHz (suitable for a 5GHz core with a 128-bit network interface), connecting a transmitter in one corner of a 22mm x 22mm chip to a receiver in the other corner—a configuration which is also used in Vantrease et al. [2008] and Nitta et al. [2012c].

3.1.5   PHOTONIC POWER REQUIREMENTS

The amount of laser power required for an on-chip optical network to operate correctly is based on a worst-case analysis of the photonic losses that will be experienced as the photons move from transmitter to receiver. The on-chip attenuation (A) sources include the waveguide, waveguide intersections, waveguide bends, and photonic vias (grating couplings), as well as attenuation due to the rings themselves.

IN DEPTH: PHOTONIC POWER CALCULATION

The required laser power per wavelength is a simple calculation of $P_{PD}10^{\frac{A}{10}}$ (where P_{PD} is the required power at the photodetector and A is the attenuation of the path in dB). Regardless of whether or not TIAs are being used to amplify photodiode currents, the required power at the photodetector (P_{PD}) is based upon two factors: the amount of energy needed to switch the photodetector from a zero to a one (refered to as E_L in the "In Depth: Photodetector Efficiency"

---

[2]Bandwidth is defined differently in the signal processing realm—in this book, the word bandwidth by itself will refer to this definition, while 3dB bandwidth will indicate the definition commonly used in elecrical communication and in the "In Depth: Microring Resonators" in Section 2.1.2.

in Section 2.3.1), and the switching frequency (f_{mod}). In order to switch the optical signal at f_{mod}, the power that reaches the photodetector P_{PD} must be greater than or equal to $E_L f_{mod}$.

The rings themselves also attenuate the signals—as was discussed in Section 2.1.3, when modulating an active (on-resonance) ring the signal deteriorates as the amount of current injected into the base increases. Increasing current injection also shifts the resonant point of the ring toward the blue, and the more the resonance needs to be shifted, the more the signal is attenuated. (Even passive microrings attenuate their on-resonance wavelengths as they transition from one waveguide to the other.) In addition, the signal deteriorates as it passes by microrings that are tuned to resonate to a different value (off-resonance rings). When the estimate for required laser power is measured at the output power from the photodetector, as is done in Ahn et al. [2009] and Nitta et al. [2012a], the photodetector is also considered an on-chip attenuation source.

Including only the on-chip attenuation sources allows one to calculate the photonic power that must enter the chip, which is what is most significant to computer architects since the amount of heat that can be removed from a chip in a cost-effective manner (also known as the TDP) is

Table 3.1: Optical loss parameters from literature

Description	Minimum	Mode	Maximum
Waveguide			
Propagation	0.1dB/cm [*]	1dB/cm [†‡§]	3.6dB/cm [¶]
Intersection	0.05dB [*ℵ]	0.05dB [*ℵ]	0.2dB [*]
90° Bend	0.00215dB [ⴳ]		0.5dB [ℜ]
Microring			
On-Resonance	0.1dB [†]	1.0dB [*ℜ]	1.5dB [*ℵ]
Off-Resonance	0.0001dB [ℵ]	0.001dB [†*]	0.1dB [*]
Coupler			
Layer-Layer (Via)	1dB [ℜ]		3dB [*]
Fiber-Chip	1dB [†‡§*]	1dB [†‡§*]	3dB [ℜ]
Photodetector			
Loss	0.1dB [*]	3dB [†‡]	3dB [†‡]
Splitter			
Loss	0.1dB [†‡]	0.2dB [*ℵℜ]	0.2dB [*ℵℜ]

[*] Koka et al. [2012].
[†] Vantrease [2010].
[‡] Ahn et al. [2009].
[§] Lipson [2005].
[¶] Pan et al. [2009].

[⋆] Pan et al. [2010].
[ℵ] Joshi et al. [2009a].
[ⴳ] Reference to Vlasov et al., IBM in Koch [2006].
[ℜ] Kirman et al. [2006].

fixed [Intel, 2004, 2011]. If one is interested in the total wall-plug power, however, the losses due to coupling, the losses in fiber-optic cables, and the efficiency of the laser should also be taken into account.[3] The efficiency of off-chip lasers is a serious concern, but it is a parameter the architect does not have control over—thus, in this book we will focus on things the architect *can* control (the on-chip power dissipation), and not the power consumed at the wall-plug. Our goal is to present the advantages and disadvantages of on-chip photonics—if the advantages are considered high enough, it will spur more research into ways to increase the efficiency of external lasers. (In fact, in Chapter 6 we will discuss some of the exciting work in on-chip lasers that is going on right now.)

The photonic path attenuation is also affected by the width of a phit. Using fewer wavelengths and switching faster (a narrower/faster link) has the advantage of requiring the photonic signal to pass through fewer off-resonance microrings and thus experience lower attenuation; however, switching at a higher frequency requires a microring with an accordingly larger 3dB bandwidth. And in order to maintain signal quality, the higher bandwidth filter center-point must be shifted further when modulating—if it is not, the signal level difference between a zero and a one could be so small that it could become indistinguishable. Since the higher bandwidth filter center-point needs to be shifted further, it will require more current injection, and thus will result in a higher signal attenuation.

The attenuation values used by different researchers for the components of optical links vary quite dramatically. Table 3.1 lists the minimum, maximum, and most frequently used values for a variety of optical parameters, taken from a sample of over a dozen sources. The table shows that in most cases there is a substantial gap between the minimum and maximum values used; what does not show up in the table is that there is also a wide range of loss numbers depending upon the specific technology that was assumed (e.g., waveguide type (channel vs. slot), or composition of waveguide (Si vs. Si_3N_4)). Some published articles provide more useful information than others; for example, in Vantrease [2010] and Ahn et al. [2009] the waveguide attenuation is separated into two values, one for single-mode waveguides and one for multi-mode waveguides. In addition, Vantrease [2010] separates the off-resonance microring attenuation into different values for adjacent and non-adjacent wavelengths (e.g., λ_1 or λ_3 passing a micoring tuned to λ_2 vs. λ_1 passing a micoring tuned to λ_3). Others, such as Joshi et al. [2009a] and Pan et al. [2010], include a non-linear waveguide propagation loss due to TPA (discussed in Section 2.2.1). Thus, designers should be very careful when choosing loss values from the literature, and if possible, architects should work with device experts to get a *consistent* set of attenuation values (whether they are optimistic and/or conservative).

---

[3]It is important to remember that if a wall-plug power comparison between optical and electrical systems is being done, the architect must not forget to account for *all* the components that contribute to the electrical system power consumption, such as DC/DC conversion and AC/DC conversion efficiencies. If this is not done, the comparison will instead wind up being between wall-plug optical and on-chip electrical power consumption.

3.1.6   ELECTRONIC POWER REQUIREMENTS

The electrical power required by a microring-based photonic link depends primarily upon four things: the power needed for trimming (we discuss this in detail in Section 5.3), microring modulation, the SERDES circuitry, and the power required to send the signal from the network interface to the microring drivers (the local transport power). Most researchers only incorporate the first two factors when calculating energy consumption, ignoring the other two. In this section we will explain the importance of including all four.

Trimming

The trimming power (also known as tuning power) is the additional power required to trim (tune) the microrings' resonance frequency. It is needed in order to overcome fabrication inaccuracies and/or to correct for environmental influences. It was shown in Nitta et al. [2011a] that system-level trimming power has a non-linear relationship with microring count, and that a power/thermal analysis of an entire design must be done to accurately estimate the required trimming power. Issues such as thermal crosstalk and the positive feedback of current injection (current injection heats the rings, heat in the rings causes red shift, requiring more current injection, etc.) make proper analysis of trimming power non-trivial. The trimming power is essentially independent of the choice of data width/frequency, and for the purposes of this discussion can be considered a static power overhead. (It is not actually constant, but it can be considered a static component because it does not change rapidly enough to be considered a dynamic component.)

Microring Modulation

The power required for microring modulation is dominated by the microring capacitance and the switching frequency. Since the microring capacitance does not change, the power required also remains somewhat constant across various data width/frequency pairings for a given target bandwidth.

Serializer-Deserializer (SERDES)

SERDES power consumption will increase if a narrow, fast link is used. In addition, the complexity of the serialization structure (whether using a serial shift register or multiplexer) increases with the degree of serialization, as well as when the frequency of switching is increased.

Local Transport

The length of the wires required to provide local transport within a node is typically ignored. However, the length of the wires required to support the microrings must be considered in the detailed power model, since the distances are large relative to the technology node. For example, a 64-bit link using a conservative ring pitch of $20\mu m$ ($10\mu m$ diameter ring, $10\mu m$ spacing between rings) has a worst-case end-to-end distance of $1280\mu m$. Even if the most aggressive ring pitch is used ($3\mu m$ diameter ring, $5\mu m$ spacing between rings) the distance is $512\mu m$, which is

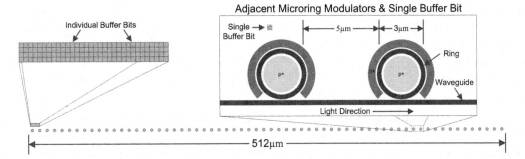

Figure 3.2: Relative sizes of electrical and photonic components. In this figure, the amount of area required by 64 microring resonators using an 8 m pitch is shown. The relative size of the electronics (assuming a 16nm technology) is included for purposes of comparison. The grey rectangle over the resonator rings represents an entire 128-bit electrical buffer. Note that the signal from the buffer to the farthest ring must travel approximately 1000 bit-cell widths. If the larger rings are used (20 m pitch), the rings would cover over 1250 m, and the signal would need to travel approximately 2500 bit-cell widths.

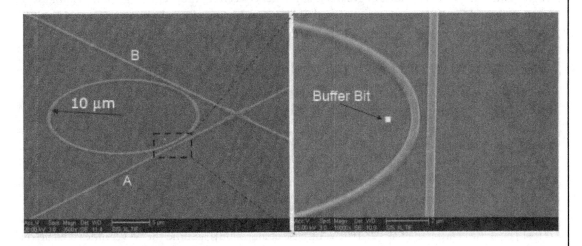

Figure 3.3: To help the reader get a sense for the size difference between the rings and the electronics, this is the same 10 m diameter microring shown in Figure 2.2 with a white square added to the picture which is the size of a single bit of a buffer in 16nm technology.

approaching the same order of magnitude as that assumed for inter-node links in an electrical mesh for a 2–3mm-wide tile.

Figure 3.2 illustrates the relative distance for 64 microrings with an 8 m ring pitch and a 16nm electrical technology point. The grey rectangle above the 64 rings is equivalent to the area

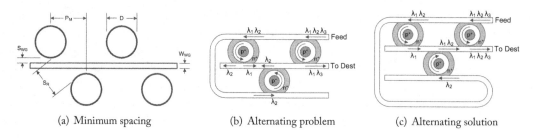

| (a) Minimum spacing | (b) Alternating problem | (c) Alternating solution |

Figure 3.4: Alternating microring spacing (a), problem (b), and complex solution (c).

needed for a 128-bit flit buffer (in a 4×32 configuration). Note the relative size of a standard cell of a D flip-flop (commonly used in buffers) compared to the size of the microring, and more significantly, the relative distance an electrical signal must travel to get from the buffer to the farthest ring. To further illustrate the relative size differences for the reader, in Figure 3.3 we have overlayed a small box that is the size of a single bit cell using 16nm technology on the photomicrograph of the actual fabricated 10μm passive microring. The distance that the electrical signals must travel will become of greater and greater concern as the technology point changes and feature sizes decrease, since (as we pointed out earlier) Moore's Law does not apply to photonics (i.e., the rings cannot get any smaller than 3μm).[4]

3.1.7 LAYOUT/IMPLEMENTATION ISSUES

We simulated a wide variety of microring layouts and found that the linear layout shown in Figure 3.2 was the most energy efficient design for both the 8μm and 20μm microring pitches. A natural question to ask is why the microrings in Figure 3.2 are laid out linearly—why can't the waveguide just weave back and forth through a grid of microrings? The problem with using a grid approach to laying out a single link is the number of tight bends that would be required. For the case of 64 microrings in an 8×8 grid, this could increase the photonic signal attenuation by as much as 2.8dB (assuming 0.1dB per 90° bend[5]), which would almost double the required photonic power. And it may not even be possible—many of the proposed on-chip photonic networks (Joshi et al. [2009a]; Pan et al. [2010]; Vantrease et al. [2008]) assume the waveguides are laid out in lanes around the chip, and putting the microrings in a grid would require a dramatic increase in the complexity of the entire network.

Furthermore, it is not possible to place microrings on both sides of the waveguide directly opposite one another, because microrings that are too close will have crosstalk/coupling issues. Figure 3.4(a) shows the best linear arrangement possible—however, for typical microring diameter

---

[4]A potential fifth factor in the electrical power consumption is the power needed by receiver amplifiers such as TIAs. Since we are assuming that TIAs will not be used in the design, we will not discuss this source in detail, but TIAs consume both static and dynamic power.

[5]A 0.1dB loss is a reasonable assumption considering that 0.32dB 90° bends have been demonstrated in Qian et al. [2006] and losses of 0.0043dB per 180° 6.5μm radii bend have been demonstrated by Vlasov et al., IBM [Koch, 2006]

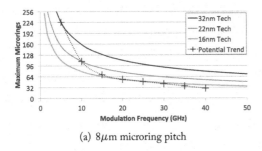

(a) 8μm microring pitch

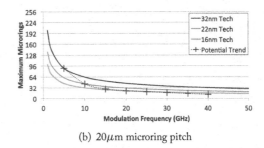

(b) 20μm microring pitch

Figure 3.5: Maximum number of microrings that can be used in a single link without using electrical repeaters, for 8μm Pitch (a) and 20μm Pitch (b), plotted vs. Switching Speed (GHz). As the technology point decreases, so does the distance that the supporting wire can carry a signal at the same speed.

and spacing values, this only decreases the total length by about 13%. Furthermore, if modulating ones, this approach has the problem shown in Figure 3.4(b). This can be solved using the approach shown in Figure 3.4(c), but it is clear that doing so increases both the area and the number of waveguide bends (and thus the signal attenuation).

Increased network complexity is also why a wider data path cannot easily be split across multiple waveguides (e.g., a 64-bit data path using 4 waveguides, each with 16 wavelengths). Designs that can accommodate the additional area requirements may benefit from splitting the data path across multiple waveguides; however, it is important to understand that increasing the number of waveguides can have a big impact on the final layout. Using Corona [Vantrease et al., 2008] as an example, if 16 wavelengths per waveguide are used instead of 64, the number of waveguides necessary will quadruple. Assuming an 8μm ring pitch and 4 passes of the serpentine, the die height would need to increase by approximately 50%, to 33.5mm. Adding waveguides is non-trivial and would increase the degrees of freedom in our analysis—instead, we will focus on varying the number of microrings in the source/destination nodes, since that should only affect the endpoints.

Electrical Wiring and Repeaters

Given the distance electrical signals will have to travel from the buffers to the rings, it may be necessary to use repeaters. Figure 3.5 shows the maximum number of microrings that can be supported without the use of repeaters on the y-axis, and the switching speed on the x-axis. The maximum distance a wire can carry a signal was calculated using the bandwidth equation from Naeemi et al. [2004] and wire technology data from ITRS 2011 [Semiconductor Industry Association, 2011]. The data points in the figure were determined by calculating the maximum length a global wire could carry a signal for a given bandwidth and target technology point. Figures 3.5(a) and 3.5(b) assume an 8μm pitch (3μm diameter, 5μm spacing) and a 20μm pitch

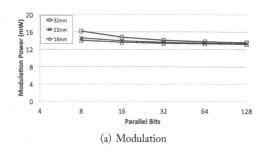

(a) Modulation

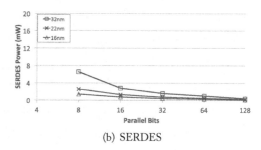

(b) SERDES

Figure 3.6: Electrical power (mW) vs. parallel bits for modulation (a) and SERDES (b). The electrical power needed for modulation increases with the switching speed because drivers must be increased in size to switch at the higher speed—the increased size means increased capacitance and higher power consumption. The SERDES structures must also increase in size with increased switching speed. The SERDES structures are not only compensating for the switching speed, but also are increasing in complexity (the number of parallel bits serialized/deserialized increases).

($10\mu m$ diameter, $10\mu m$ spacing), respectively. These two pitches were chosen because they represent theoretically possible and currently practical design points that have appeared in the related literature. The figures show that as the switching speed is increased and the technology is scaled, fewer and fewer microrings can be reached without requiring repeaters (shown by the line marked *potential trend*). Adding repeaters is a concern, because while they do allow a signal to be sent further, repeaters will also either add latency or increase power consumption. And if the choice is to use repeaters in a way that does not increase latency, eventually not even repeaters will work, because the sub links between the repeaters will be unable to switch fast enough to carry the higher sublink bandwidth.

3.1.8   WIDE AND SLOW OR NARROW AND FAST?

The complexity and power consumption of the SERDES favors the use of a wider, slower link, but such a link requires many rings, and the cost of getting electrical signals from the buffers to the rings is significant and argues in favor of a narrower, faster link. In other words, the architect must come to grips with the fact that as the electrical geometries shrink, the bandwidth of a photonic link may be limited in ways that were not originally apparent. In order to evaluate this situation more fully and attempt to determine what combinations of switching speeds and phit widths are best for given technology points, we ran a range of simulations, which we will discuss next.

As described previously, the amount of electrical power used has four main sources—SERDES, modulation, transport, and trimming/tuning. The first three are dynamic in nature, while the *trimming* component is a source of power consumption which is highly temperature dependent and more static in nature (the trimming power will vary over time, but not nearly as fast as the other three do). It is evaluated by calculating the total dynamic power used and the

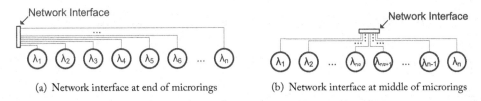

(a) Network interface at end of microrings   (b) Network interface at middle of microrings

Figure 3.7: Microring wiring with network interface at the end (a) and middle (b) of a row. The network interface is the electrical connection to the link, and is assumed to be flit-width bits in size. If the interface is at the end of the row of microrings, the worst-case wire length is double that of the situation where the network interface is at the middle. In addition, the total amount of wire is greater if the network interface is at the end.

area of the various components, and feeding these values into a thermal simulator that takes into account temperature-dependent static electrical power consumption. Trimming is discussed in more detail in Section 5.3.

Figure 3.6 shows the electrical power required in mW on the y-axis vs. the number of parallel bits on the x-axis for the modulation of the microrings (Figure 3.6(a)) and for SERDES (Figure 3.6(b)) in a 640Gbps link at various technology nodes. The figure shows that the power required for modulation goes down as the number of parallel bits increases, since the switching speed decreases. The modulation power required is also reduced at each successive technology node, due to the reduced capacitance of the driver circuit; however, the relatively fixed capacitance of the microring resonator limits the amount the modulator power can be reduced.

Figure 3.6(b) shows the power required by the SERDES (modeled as a pass-gate multiplexer for serialization and a single serial shift register shifting through an enable bit for deserialization) decreases as the number of parallel bits used increases, since both the switching speed and the SERDES structure capacitance are simultaneously decreasing. Each successive technology node also shows a reduction in power. The power required for both modulation and SERDES clearly favor the use of a higher number of parallel bits, as we have pointed out before.

The flit buffer (seen in Figure 3.2) serves as the interface between the core and the network. In our experiments we placed the buffer either at the end of the row of microring resonators or in middle (shown in Figure 3.7.) These two configurations were chosen because "End" has the longest worst-case wiring, while "Mid" has the shortest. (We felt it was necessary to look at both configurations, even though "Mid" is clearly preferable, since floorplan issues might force the architect to use the "End" configuration in certain circumstances.) Figure 3.8 shows the electrical power required to move the data from the buffer to the rings (the local transport power) in mW on the y-axis and the number of parallel bits used on the x-axis. Since "End" and "Mid" are the worst and best cases for transport wire length, we chose to use conservative wire projections for "End" and optimistic wire projections for "Mid" in our experiments. These two configurations are labeled "Worst and" "Best" in the results that follow.

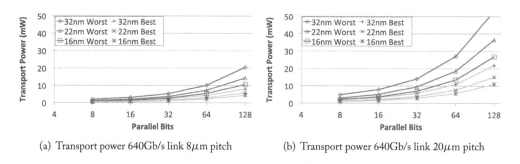

(a) Transport power 640Gb/s link 8μm pitch

(b) Transport power 640Gb/s link 20μm pitch

Figure 3.8: Electrical power (mW) vs. parallel bits for local transport at 640Gb/s link for 8μm pitch (a) and 20μm pitch (b).

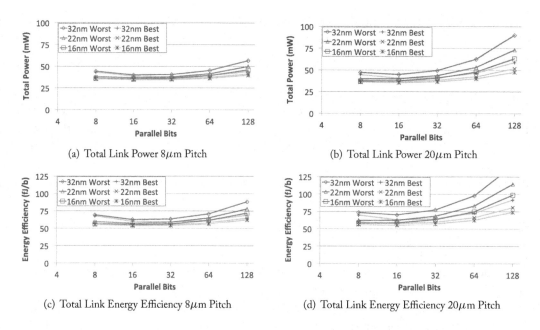

(a) Total Link Power 8μm Pitch

(b) Total Link Power 20μm Pitch

(c) Total Link Energy Efficiency 8μm Pitch

(d) Total Link Energy Efficiency 20μm Pitch

Figure 3.9: Total Link Power (mW) (a) and (b) and Total Energy Efficiency (fJ/b) (c) and (d) vs. Parallel Bits. The power and energy efficiency values assume a link with 100% utilization. The figures show that in general, the larger the microring pitch, the larger the difference between the optimal and the other configurations.

The wire to each microring was properly sized (local, semi-global, global) using the technology data from ITRS 2011 [Semiconductor Industry Association, 2011]. Some of the more extreme cases (e.g., "Worst" for 64 & 128 bits) required repeaters in order to provide the proper

Table 3.2: Parallel bits & convergence point for most energy efficient configuration based on link utilization

Tech Point	8 μm	20 μm	Tech Point	8 μm	20 μm
32nm "Best"	16 < 39% > 32	16	32nm "Worst"	16	8
22nm "Best"	16	16	22nm "Worst"	16	8
16nm "Best"	16	16	16nm "Worst"	16	8

bandwidth to some of the furthest microrings, and the repeater power is included for those cases. The repeaters were sized to maintain overall link latency, not to maximize energy efficiency.

As one would expect, Figure 3.8 shows that the power needed increases with the increase in wire length—the "Worst" values require higher power than the corresponding "Best" values. The total capacitance of the wires is roughly bound by $\frac{N(N+1)}{2}$, where N is the number of parallel bits, leading one to expect the power to increase quadratically; however, as the number of parallel bits increases the switching rate decreases, resulting in an approximately linear growth. As discussed in the previous section, the number of microrings that can be supported without needing repeaters decreases with each successive technology point, as does the transport power (since the capacitance of the wires decreases).

3.1.9 TOTAL POWER

Figure 3.9 shows the total link power in mW (Figures 3.9(a) and 3.9(b)) and the total link energy efficiency in fJ/b (Figures 3.9(c) and 3.9(d)) for links assuming both the "Best" and "Worst" cases. 16 parallel bits is the most energy efficient across all the configurations with a 8μm microring pitch, while the most energy efficient configuration with a 20μm ring pitch is 16-bits and 8-bits for the "Best" and "Worst" case respectively, regardless of technology point. Note that the figures also show that there is a limit to how narrow the bus width should be—8 bits is only the most efficient for the "Worst" case of the 20μm ring pitch. In Nitta et al. [2012b], even narrower links (as low as 2 and 4 bits) are shown to be substantially less efficient because of the increased SERDES power requirements.

Impact of Utilization

The power and energy efficiency values shown in Figures 3.9 assume a 100% link utilization, which is unrealistically high in the steady state. In fact, the actual energy efficiency is *highly* dependent on the link utilization, because photonic links require a large amount of static power. Remember, the off-chip laser has to be sized to be able to provide enough power for the link to work in the worst-case situation (100% utilization), but that power is not turned off when the link is idle. It simply comes on-chip and is not used, making it a significant static overhead. (This is also why there is such interest in on-chip lasers—it may be possible to turn off an on-chip laser when a link is idle, providing a substantial energy savings.)

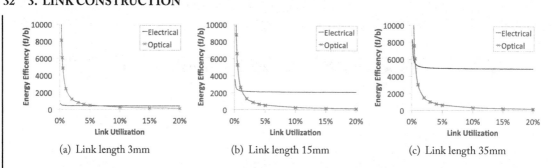

(a) Link length 3mm (b) Link length 15mm (c) Link length 35mm

Figure 3.10: Energy efficiency (fJ/b) vs. link utilization using 16nm and 640Gb/s for short (a), medium (b), and long distance (c). When the link length is short electrical links are more efficient until the link utilization rate becomes approximately 8% (the crossover point). For medium length links, the crossover point is approximately 1%, and for long links it is 0.5%.

The energy efficiency of each of the configurations was analyzed while varying the link utilization, and the results are presented in Table 3.2. For the one entry that has more than a single value, the percentage in the middle is the link utilization point where both configurations are equally efficient, and the two numbers on the left and right of it are the number of parallel bits that are most efficient below and above that utilization rate, respectively. While it is not shown in Table 3.2, the expanded results in Nitta et al. [2012b] show a trend that as the technology shrinks, the static power becomes a larger portion of the total link power and the utilization point where the two configurations are equal increases.

Impact of Link Length

Researchers have shown that below a certain distance (known as the partition length L_{part}) it is better to use conventional electrical interconnects than photonic (or even equalized electrical) ones [Beausoleil et al., 2008; Kim and Stojanović, 2010; Naeemi et al., 2004]. However, we have shown that the energy efficiency of photonic links is highly dependent upon utilization, a fact which is not included in the equation used to calculate L_{part}. Therefore, we decided to investigate the relationship between link utilization and link length.

IN DEPTH: PARTITION LENGTH

The term partition length (L_{part}) refers to the distance beyond which information can be more efficiently transported by photons rather than electrons, and is described by the following equation: $L_{part} = \frac{W_{min}}{N} \sqrt{\frac{K_0}{f_{max}}}$, where W_{min} is the minimum waveguide width, K_0 is a constant that depends on the material and the conductivity mechanism, f_{max} is the maximum modulation frequency, and N is the number of wavelengths per waveguide or the WDM factor. For example, at

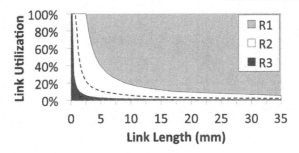

Figure 3.11: Transition point where optical and electrical link costs are the same. In the upper right part of the figure optical is the most efficient, in the lower left electrical is.

10GHz, the partition length for a waveguide width of 1μm with a WDM factor of 10 is about 0.5mm.

Figure 3.10 presents the results using a 16nm technology point and 640Gb/s bandwidth for three different link lengths—3mm (representative of communication between adjacent nodes), 15mm (communication between nodes in the same row/column), and 35mm (corner-to-corner worst-case length). We chose to show the results for this technology point because it is the most aggressive—other configurations have the same basic shape as this graph, with the only difference being the electrical line is at a different (higher) point.

Several things are clear from looking at Figure 3.10. For one, optical interconnects are always more efficient than electrical interconnects when the utilization is high. However, if the link length is small, the difference is small as well, and optical links may not be worth pursuing. On the other hand, optical interconnects are substantially more efficient at high utilization when the distances are large.

Figure 3.10 also shows that the average link utilization must be extremely low for electrical interconnects to be the most energy efficient. In Figure 3.11 we plot the link length vs. the link utilization to show the relationship between the two. There are three regions shown—in the R1 region, optical interconnect efficiency is higher, while in the R3 region conventional electrical links are more efficient. The R2 region that lies in between is the range of values where the most efficient is dependent upon a variety of factors—the technology point, the optical/electrical assumptions, the target bandwidth, etc. The dashed line in the R2 region is where electrical and optical interconnets/links are equally efficient, assuming the most aggressive and optimistic electrical and optical assumptions.

Caveats

It should be noted that the results in Figure 3.10 are for conventional electrical interconnects, and do not include more advanced schemes such as equalized interconnects [Joshi et al., 2009b]. However, given the limited data points available from the literature and the assumptions made

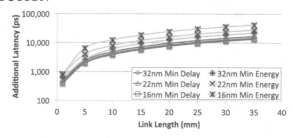

Figure 3.12: Electrical latencies as link length changes. Min Energy indicates delay when electronics are sized to minimize power consumption while maintaining bandwidth, while Min Delay indicates delay when electronics are sized to minimize delay.

in this work, we expect that such advanced electrical signaling schemes would reside in the R2 region, and could potentially expand it somewhat.

It is also worth noting that electrical interconnects cannot match the low flit latency provided by photonic links. The electrical interconnects analyzed had an additional 1–42ns flit latency due to the number of repeaters necessary—the repeaters were added to support the desired bandwidth, not to reduce link latency. In Figure 3.12 we plot the impact of link length on latency, and show that as link length increases, the electronic latencies increase as well. The bottom 3 lines in the figure represent the delays experienced by the three different technology points when energy usage was ignored and the repeaters and spacing were sized purely to minimize delay. Even in this case there is an additional 0.6–17ns flit latency. The top 3 lines are what happens to the delays when the repeaters were sized to make them most efficient. As one would expect, the figure shows that as the technology point moves toward smaller and smaller features sizes, the link delays incurred as the link length increases become larger and larger.

Finally, it is important to keep in mind that the design space when dealing with photonic links is extremely large, and the results are sensitive to the assumptions made about underlying photonics and electronics. Readers interested in learning more about this topic should see Georgas et al. [2011] and Ophir et al. [2013], which describe the trade-offs in photonic link design using slightly different assumptions.

CHAPTER 4

On-Chip Photonic Networks

In this chapter we will look at how to build an on-chip photonic interconnection network using the links described in Chapter 3. We begin with an overview of how the basic aspects of an on-chip network, such as topology, routing, arbitration, switching, and flow control are impacted by the use of photonics. Once we have discussed these various aspects, we will present an overview of the on-chip photonic networks that have been proposed by different research groups, with an emphasis on comparing how they deal with the different photonic networking challenges.

4.1 PHOTONIC NETWORK DESIGN CHALLENGES

In this section we will describe how some of the unique characteristics of photonics impact the design and optimization of a network on a chip.

4.1.1 BUFFERING

Packet-switched networks are based on the store-and-forward paradigm—a packet of data is stored (buffered) until the resources required to send it to its next destination are available. Routing algorithms, flow control mechanisms, and arbitration protocols all take advantage of the fact that it is relatively easy and inexpensive to store a packet in the electronic domain, because storage structures like registers and FIFOs can be implemented efficiently and scale with technology. Unfortunately, light cannot be stored so easily, because a photon is a massless, chargeless particle traveling at the speed of light—at best one can build a rudimentary delay line memory, which provides nowhere near the flexibility and functionality of a RAM. At this point the only viable approach in the arena of on-chip photonics is to convert photons back to electrons and route and store them using conventional electronics, which not only introduces additional complexity but also results in additional power, area, and latency penalties.

4.1.2 TOPOLOGY

The inability to store photons has a direct impact on the selection of a topology. Simple topologies like meshes and tori with only near neighbor connections are easy to implement, but they require buffering at the intermediate nodes, which (as we mentioned above) increases complexity. On the other hand, the ability of photonics to communicate across long distances without needing repeaters and without compromising the data rate favors buses, crossbars, and fully-connected topologies. These topologies require only a single hop between the source and the destination and hence reduce latency as well. (However, fully connected topologies increase the number of waveguides and the number of transmitters and receivers, and therefore impact scalability.)

4.1.3 ARBITRATION AND FLOW CONTROL

In the electronic realm, contention resolution (arbitration) is simple—the packets are held in a buffer until an arbiter gives one of the packets exclusive access to the shared resource. However, if the arriving packets are photonic in nature, contention resolution becomes much more challenging. The simplest solution is to convert the packets to the electronic domain and use standard contention resolution mechanisms, but this approach introduces both complexity and additional latency in the critical path of the packets.

Another approach is to do arbitration in the optical domain, as is done in the Corona [Vantrease et al., 2008] on-chip photonic network. It is also possible to first establish a path between the sender and receiver using electrical signaling, and then stream packets (essentially use circuit-switching). This is the approach taken by the authors in Shacham and Bergman [2007], Shacham et al. [2007a], and Shacham et al. [2007b]. Or, one can avoid the need for arbitration altogether by using a directly connected network, which is the approach put forth in DCAF [Nitta et al., 2012c].

However, even in networks which do not require arbitration, there will be the need for flow control. Flow control is required in a network to prevent a sender from overwhelming a slow receiver, and relies on the ability to disseminate the occupancy status of buffers at a receiver to all potential senders. In an electrical network this is a problem, because the performance of electrical wires does not scale with distance. Photonics, on the other hand, has an advantage here—in fact, one could imagine using photonics just for flow control in an otherwise all-electrical network-on-chip.

Traditionally, a credit-based scheme such as that described in Dally and Towles [2004] is used to implement flow control in an electrical network. However, the higher bandwidth and lower energy per bit available when using photonics could enable simpler schemes, such as re-transmitting based on a Negative AcKnowledge (NAK) signal, as described in Cianchetti et al. [2009]. If the cost of flow control is low enough, it could potentially enable a reduction in the amount of buffering required in a network (with the resultant advantage of possibly lower power and area).

It is important to remember, though, that while photonics may be better than electronics at disseminating the global information needed to implement flow control, the power required may be significantly higher as well. This is particularly true if an external fixed-size laser is used, because (as described in Chapter 2) the laser power has to be large enough to satisfy worst-case demands. The use of on-chip lasers that can be turned on only when needed might make this a less serious issue.

4.1.4   ELECTRICAL/OPTICAL CODESIGN

Electrical/optical codesign involves making sure that electronics are used when electronics are most appropriate (when the communication distance is very short, for example) and photonics are used when photonics have an advantage (e.g., when the distance is longer). This can lead to network designs which have a two-level hierarchy—an electronic network (a mesh or some other suitable topology) to connect processors that are in close physical proximity, and a photonic network (a ring or crossbar, for example) to connect the processor groups. The exact size of the groups and how they are connected to the photonic network, as well as the number of links there should be, the bandwidth, the expected utilization of the link, etc. presents an interesting set of questions and defines the design space for the photonic network architect.

Another example of electrical/optical codesign is to augment an electronic on-chip network with optical instead of electrical express channels.[1] The use of optical express channels is attractive because they are relatively straightforward to design and have the least impact on the manufacturability (and hence the cost) of a chip that incorporates photonics. The partition length (described in Chapter 3) can be used as rule of thumb to help decide the location and the bandwidth of the optical express channels.

4.1.5   LATENCY

Just as the execution time equation (Instruction Count x Clock Cycles Per Instruction x Clock Period) provides insight into the various tradeoffs in the performance of a uniprocessor, the network latency equation shown below can be used to highlight the differences and tradeoffs in an on-chip photonic network. The latency equation is

$$Latency = T_{send} + (d + 1)T_{prop} + (T_r + T_a + T_s)d + \frac{Pktsize}{BW} + T_{rec},$$

where T_{send} and T_{rec} are the sending and the receiving overheads, d is the number of hops between the source and the destination, T_{prop} is the propagation delay through the waveguide, T_r, T_a, T_s are the routing, arbitration, and switching delays through the router, $Pktsize$ is the size of the packet, and BW is the link bandwidth.[2]

As noted in the topology-related discussion above, photonics favors networks with smaller hop counts (smaller values of d) because less intermediate buffering is required. That's a significant advantage for photonics over electronics, since the parameter d is a multiplier for both the propagation and the router delay. With single hop networks like crossbars, T_r is not significant

---

[1]Express channels are direct connections between nodes that are relatively far apart, in order to reduce congestion and average latency in an electrical on-chip network (at the expense of requiring more complex routing protocols).

[2]Even when using DWDM, the number of wavelengths per waveguide will be significantly smaller than the flit size. In fact, as we explained in Chapter 3, due to energy efficiency concerns it might be appropriate to use narrower data paths in photonic links. So the 128- or 256-bit flits have to be squeezed into the narrower photonic data paths, which involves a parallel-to-serial conversion at the transmitter and a corresponding serial-to-parallel conversion at the receiver. This overhead has to be added to the traditional sending and receiving overhead.

but T_a is still important. As noted previously, the lack of buffering makes contention resolution in photonics more complex and hence the value of T_a may be higher. T_s is typically a small number when using microrings in switches, and largely independent of the size of the switch—whether it is a 4x4, 16x16 or 64x64 crossbar T_s does not change much when using photonics, but it could become quite significant in an electronic implementation. This means that photonic networks can be much flatter (i.e., they do not require as many layers in a hierarchy) which in turn reduces d.

Clearly, link BW is higher for photonic networks (in fact, that's one of the primary motivations for considering photonic networks in the first place), with 10Gb/s achievable now and 20–40Gb/s possible in the near and long term. However, if $Pktsize$ is small the bandwidth advantages of photonics may be lessened, since T_r and T_a may be dominant factors—in other words, in order to take full advantage of photonics it may be necessary to use larger packet sizes. This means that perhaps packet aggregation (as described in O'Mahony et al. [2001]) and/or traffic grooming (as presented in Cinkler [2003]), two techniques that are used in general-purpose telecommunication networks, should be considered for use in photonic on-chip networks.

In fact, the packet size argument can be used to make a case for circuit-switched photonic networks. Circuit-switching is easier, since some of the problems faced by photonics (lack of buffering and difficulty of contention resolution) can be mitigated by the fact that they only have to be dealt with on a per-circuit basis, instead of per packet. This allows photonics to spend more time do what it is good at—providing high bandwidth data transfer at a low energy per bit. The work described in Shacham and Bergman [2007] and Shacham et al. [2007a,b] is an example of this type of thinking, and is also the motivation for the current efforts to use photonics in datacenter networks [Farrington et al., 2010; Vahdat, 2012; Vahdat et al., 2011].

4.2   CASE STUDIES OF ON-CHIP PHOTONIC NETWORKS

Many microring-based on-chip networks have been proposed in the literature. In this section we will present an overview of six different designs, chosen because they all take different approaches to dealing with the various challenges and because there is enough published information about them to be able to discuss them. For each proposal, we will explain 5 things: the topology, the approach to arbitration/flow control, how it deals with buffering, the electrical/optical partitioning, and the performance of the proposed network. We will conclude the overview of each proposal with a brief discussion.

4.2.1   CORONA

Topology: The Corona design is a 256 core multiprocessor which utilizes a 256-bit optical crossbar operating at 10GHz (double clocked at 5GHz) to interconnect 64 four-core clusters (nodes); since within each cluster the cores are connected electrically, the Corona network is a clear example of a two-level hierarchy. The optical crossbar is implemented using four dedicated photonic bus waveguides per destination, in order to provide the 256-bit data path [Vantrease et al., 2008]. The waveguides in Corona carry 64 bits of data (shown in Figure 4.1(b)) and are laid out

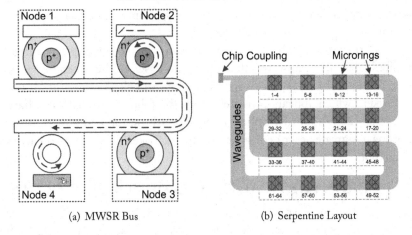

(a) MWSR Bus   (b) Serpentine Layout

Figure 4.1: A single-bit MWSR bus (a) for destination node 4 with node 2 transmitting. Wavelength is removed to dead-end drop to create a zero. Serpentine layout (b) of waveguides similar to Corona optical crossbar.

in a serpentine fashion in order to pass by every cluster. Using a serpentine layout allows each cluster to be connected, while simultaneously avoiding waveguide crossings. Each output of the photonic crossbar is actually implemented as a Multiple Writer Single Reader (MWSR) bus, so while logically it is a crossbar, physically it is a photonic bus. Figure 4.1(a) shows an example of a four-cluster single-bit MWSR bus similar to the configuration used in Corona; in this figure, Node 2 is transmitting a signal to Node 4 by modulating zeros. Note that any of the three clusters could transmit to Node 4 using the same waveguide.

Arbitration/Flow Control: Since all sources share the MWSR bus, arbitration is required in order to coordinate access to Corona's optical crossbar. Corona combines arbitration and flow control into a single, all-optical token-based mechanism, which is implemented using additional buses, which are slightly different than the MWSR buses used for the crossbar data paths. The arbitration/flow control buses allow each node to remove (consume) a token, and also to inject a new token onto the bus.

The token bus is constructed using two photonic waveguides, one that is used as a power feed and the other that carries the photonic tokens. Between the two waveguides are active microrings, which are used by each node to inject new tokens (modulating ones) onto the token waveguide. Connected to the token waveguide, opposite the token-injecting microrings, are a set of active microring resonators that are connected to photodetectors and are used to do token consumption. In order to consume a token these microrings are activated, and once the photonic token arrives at the microring resonators, it is removed from the token waveguide. Since the token is composed of light, it can be consumed completely, guaranteeing that only a single node will acquire arbitration.

The actual arbitration mechanism used in the original Corona paper, Vantrease et al. [2008], was not explicitly described; however, the authors do describe the token bus and state that, due to the nature of the protocol, a processor can wait up to 8 clock cycles (at 5 GHz) to receive an uncontested token, which appears consistent with the follow-on token-based arbitration work. In Vantrease et al. [2009], four all-optical integrated arbitration and flow control protocols used with the Corona design are described: Token Channel, Token Channel with Fast Forward, Token Slot, and Fair Slot.

IN DEPTH: CORONA OPTICAL FLOW CONTROL PROTOCOLS

The Token Channel protocol works as follows: the destination node sends out a token containing the amount of buffer space available. A node that wants to send to that destination enables its microrings to remove the token from the arbitration bus, and then transmits its flits. When it has no more data (or there is no space available at the destination) the updated token is injected back onto the arbitration bus. This prevents multiple senders from writing to the data bus at the same time, and prevents any sender from overwhelming the receiver; however, this technique does favor the nodes immediately downstream from the destination node and can potentially lead to starvation.

The Token Slot protocol attempts to simplify the process by reducing the number of wavelenths required for arbitration; here, a light pulse on a single wavelength indicates the presence of an available buffer slot. The tokens can be consumed as needed, and as soon as a buffer is freed, a new token can be issued. The downside of the Token Slot (like the Token Channel) is that the nodes upstream are favored and can starve out nodes downstream.

The Token Channel with Fast Forward and Fair Slot protocols both address the issue of starvation. The Token Channel with Fast Forward adds a Fast Forward channel and a mechanism to replenish an empty token, and the fully replenished token can skip upstream nodes so that downstream nodes cannot continually be starved. The Fair Slot protocol broadcasts hungry messages by those nodes that are being starved, and was shown to outperform the Token Channel with Fast Forward protocol; however, the power requirements of a broadcast waveguide should not be overlooked, and may be the deciding factor in favoring one protocol over another.

Buffering: Corona requires buffering on both the transmit and receive sides. The receive side of Corona requires only a single receive FIFO since the communication channel is arbitrated for, and only a single flit can be received at a time. If the Token Channel protocol is used, it is likely that the maximum value of the token will match the size of the receive FIFO. On the transmit side, a single shared buffer would not be feasible, since multiple flits can be simultaneously transmitted to different destinations—in order for the transmit buffers to be shared, an electrical crossbar would be required to connect the buffers to the transmitters. One of the most practical designs would be to have a single packet-sized (or possibly flit-sized) buffer for each destination.

Electrical/Optical Partitioning: The on-chip network proposed in Vantrease et al. [2008] electrically clusters four cores around each node of the photonic crossbar. Arbitration, flow control, and data transfer are all handled optically on the photonic crossbar.

Performance: The results presented in Vantrease et al. [2008] for synthetic benchmarks and scaled versions of SPLASH-2 benchmarks showed that by using optically connected memories and the optical crossbar, system performance could be improved between two and six times on memory-intensive workloads when compared to systems with purely electrical interconnects, while also significantly reducing the interconnect power.

Discussion: Corona was one of the first large-scale on-chip nanophotonic networks to appear in the literature, and because so many details have been published [Ahn et al., 2009; Vantrease et al., 2008, 2009; Vantrease, 2010], it has proven to be an excellent design to compare against. The proposed Corona network utilizes photonics for communication to main memory as well, although in the above description we focused only on the optical crossbar design. Interestingly, the authors felt that the off-chip path was the most significant aspect of this proposal.

Even though the results in Vantrease et al. [2008] showed significant power savings, one drawback of the MWSR approach is the number of off-resonance microrings the photons must pass in order to reach the receivers at the final destination. The MWSR approach is only feasible when the off-resonance attenuation is extremely small, or the number of microrings that must be passed is small (either there are not very many clusters being connected, or the number of wavelengths used is small).

4.2.2   PHASTLANE

Topology: In 2006, researchers at Cornell [Kirman et al., 2006] proposed one of the first uses of microring resonators in an on-chip network, using photonics to create a bus that formed a ring around the chip to connect clusters of processors. Three years later, the same research group proposed Phastlane [Cianchetti et al., 2009], a hybrid opto-electronic on-chip network that uses a low-complexity nanophotonic crossbar, and is supported by an electrical network for buffering and arbitration. Phastlane uses predecoded source routing to implement an 8x8 mesh, where control signals are sent optically to do the N,S,E,W routing. The Phastlane approach allows packets to traverse several hops without the need for store-and-forwarding at each routing point, or the need to arbitrate for a complete data path prior to data transmission. Each switch in Phastlane receives five control signals: straight, left, right, local, and multicast. Figure 4.2 illustrates the optical switch that is used in Phastlane; in this figure a packet entering from the S wants to go out the N port at the same time that a packet entering from the E wants to use the same N output. When the control signals reach a routing point, the appropriate set of microrings are enabled to switch the wavelengths (including control signals for subsequent routing points) in the desired direction. Two waveguides carry the incoming control signals, and when signals on those waveguides reach a switch, one set of signals is physically transfered to the other outgoing

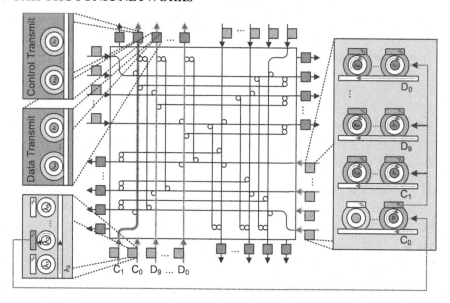

Figure 4.2: Phastlane optical switch. Incoming packet from the S port takes precedence over the incoming packet on the E input port that also desires to exit the N port. The incoming blocked packet on the E input port is received and buffered electronically until the N output port becomes available. (See Cianchetti et al. [2009] for original version of figure.)

waveguide, and the signals on the other waveguide are converted to electrical signals in order to support the desired routing. (Further details can be found in Cianchetti et al. [2009].)

Arbitration/Flow Control: A fixed-priority arbitration scheme is used at each Phastlane switch, and is implemented using a combination of both electrical and optical control signals. As mentioned above, when the optical control signals reach a switch, they are converted into electrical signals to enable/disable the appropriate microrings to prevent a packet collision.

Phastlane implements an Automatic Repeat reQuest (ARQ) based flow-control scheme that utilizes a NAK to realize a Stop-And-Wait (SAW) protocol. The source stops and waits after a transmission in a SAW ARQ protocol (thus it's name), until it receives an ACKnowledgement (ACK), either explicitly or by not receiving a NAK within a specified timeframe. It is assumed in Phastlane that, if the NAK is not received, the packet has been successfully received and buffered; therefore an explicit ACK is not sent. Since the predecoded routing control signals are enabling the data path, a return path has already been constructed for the signaling of the NAK.

Buffering: If there is contention for the outgoing path, all but one of the incoming packets is electrically buffered (as is shown happening to the packet entering from the E port in Figure 4.2).

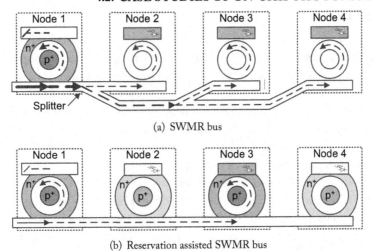

(a) SWMR bus

(b) Reservation assisted SWMR bus

Figure 4.3: Two single-bit SWMR buses, one that is implemented using splitters (a), and one that is implemented using active microrings and requires reservation assistance (b).

Packets are dropped if the desired path is contested and there is insufficient buffer space; the source node is signalled in this case using a NAK.

Electrical/Optical Partitioning: The switching and arbitration at each Phastlane switch is implemented using a combination of electrical and optical control signals. The control signals are sent between switches optically, and, if necessary, converted to electrical signals at a given switch in order to perform switching and arbitration. Some of the control signals that are converted to the electrical domain are needed later in the packet path and are therefore frequency translated ($\lambda_{35}...\lambda_6 \to \lambda_{30}...\lambda_1$) and placed back onto the outgoing waveguide.

Performance: Phastlane achieved approximately 5–10x lower latency than the comparison baseline electrical networks on synthetic traffic patterns, while also providing slightly better saturation bandwidth. The analysis of Phastlane also showed that the network achieved double the network performance of the electrical baseline on SPLASH-2 benchmarks, while consuming 80% less power.

Discussion: One downside of a NAK-only SAW protocol is that it will not protect against errors in the feedback link, since it is impossible to distinguish between a lost NAK and a NAK that was never sent; however, this protocol could easily be extended to provide reliable communication by changing the NAK to an ACK and retransmitting if the ACK is not received.

4.2.3 FIREFLY

Topology: Firefly, described by Pan et al. [2009], is another hybrid opto-electronic network that uses an electrical network for intra-cluster communication and a nanophotonic crossbar for inter-cluster communication. The Firefly architecture implements its optical crossbar with Single Writer Multiple Reader (SWMR) buses (one for each source node). In Pan et al. [2009], they describe two approaches, a splitter-based SWMR bus and a more energy-efficient SWMR bus called a reservation-assisted SWMR bus. Figure 4.3(a) shows a splitter-based, single-bit, four-node SWMR bus, which provides a dedicated transmission channel for Node 1. All the nodes receive the same data, but only the desired destination processes the received packet, making this essentially a broadcast channel. Since packets are broadcast to all destinations, this approach does not require arbitration, but does require significant photonic power. An example of a single-bit, four-node, reservation assisted SWMR bus can be seen in Figure 4.3(b). In this approach non-destination nodes turn off their receive microrings, allowing the packet to be received by the desired destination only. The packet destination node is the only one to enable its microrings (in this case Node 3), which allows the wavelengths to pass the upstream nodes (Node 2), and to be removed before reaching the downstream nodes (Node 4).

The reservation-assisted SWMR-bus approach consumes significantly less photonic power than the splitter version because it requires only enough photonic power to send to a single receiver, instead of needing enough power to send to every possible receiver—however, it requires a reservation network in order to coordinate the enabling/disabling of the receiving microrings. The reservation network is implemented using a splitter-based SWMR in which the head flit tells the destination to enable its receiving mirorings. Since this reservation bus is essentially a broadcast network, it avoids the need for global arbitration by providing a narrow, dedicated broadcast channel for every node.

Arbitration/Flow Control: Global arbitration is not needed in Firefly, since a splitter-based SWMR is used either for the entire data path, or the head flit (in the case of the reservation assisted SWMR). In Pan et al. [2009], the authors state that Firefly uses credit-based flow control for all of the channels. The credits are sent back upstream using piggybacking, and, since there are multiple buffers at each receiver, the credit information does not need to be global. There is only a single sender for each receive buffer.

Buffering: Buffering is needed for each potential source at the receive side, due to the fact that a node in Firefly can receive from multiple sources simultaneously. Since each node can only send to a single destination at a time, a single transmit FIFO will suffice.

Electrical/Optical Partitioning: Firefly uses an electrical network for intra-cluster communication, and an optical network for inter-cluster communication. Firefly implements the inter-cluster data and flow control mechanism all optically, and no global arbitration is necessary.

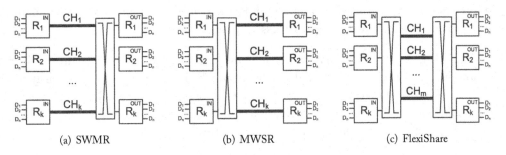

Figure 4.4: Three *n*-bit wide radix-*k* nanophotonic crossbars: SWMR (a), MWSR (b), and *M* data channel FlexiShare (c). Each R$_i$ router is split into an IN and OUT section in the figure, but would be implemenated as a single physical router. (See Pan et al. [2010] for the original version of the figure.)

Performance: In Pan et al. [2009], the authors state that Firefly improved performance by up to 57% and the Energy Delay Product (EDP) by 51% for synthetic traffic patterns when compared to an electrical concentrated mesh, and improved performance by 54% and EDP by 38% for traffic patterns with locality when compared to an all-optical crossbar.

Discussion: One downside of this approach is that the reservation channels could potentially require as much photonic power as the reservation-assisted SWMR data channels (though clearly less than using the splitter-based SWMR buses for the data channels). In addition, the timing required to coordinate the head flit with the following data flits may also be difficult to achieve, since the head flit must be received, processed and the appropriate microrings must be enabled before the arrival of the data flits.

4.2.4 FLEXISHARE

Topology: FlexiShare [Pan et al., 2010] is a flexible photonic crossbar that is a combination of an MWSR and an SWMR design. FlexiShare decouples the number of communication channels from the number of nodes, in an attempt to reduce the required photonic power and component complexity while still maintaining performance. The MWSR sending side allows any of the nodes to communicate to any of the *M* data channels, each of which is connected to an SWMR bus that allows communication to the destination nodes. Figure 4.4 illustrates three nanophotonic crossbar designs: SWMR (Figure 4.4(a)), MWSR (Figure 4.4(b)), and FlexiShare (Figure 4.4(c)).

Arbitration/Flow Control: FlexiShare employs a token stream for arbitration and uses credit sharing, adopting the reservation-assisted scheme from Firefly. This token-stream protocol differs from the token-ring arbitration discussed in Vantrease et al. [2008] and Pan et al. [2009], but is similar to the Token Slot discussed in Vantrease et al. [2009]. The authors of FlexiShare extended the arbitration work in Pan et al. [2011], where they describe FeatherWeight, an all-optical abritration scheme that exploits the benefits of nanophotonics to simultaneously provide

Quality of Service (QoS) as well as arbitration. FeatherWeight leverages a global communication path to provide a weighted min-max fairness scheme.

Buffering: Like Firefly, FlexiShare will require buffering for each potential source at the receive side; however, the number of potential sources from which a node can simultaneously receive is M (the number of data channels), not the number of nodes. Unlike Firefly, each node in FlexiShare could potentially send simultaneously to multiple destinations—therefore, M FIFO will be needed on the transmit side (instead of a single FIFO).

Electrical/Optical Partitioning: In FlexiShare, arbitration, flow control, and data transfer is handled optically. The routing of packets is handled electrically, and is stated to be more complex than in FireFly.

Performance: The results presented in Pan et al. [2010] show that FlexiShare provides comparable performance to networks that have the same number of data channels as nodes, because the M data channels in FlexiShare maintain a higher utilization. FlexiShare is also able to improve throughput by 5.5x when compared to a token-ring arbitration network, while reducing overall power consumption between 27% and 72%. The results on traces from SPLASH-2 benchmarks showed that FlexiShare could reduce the execution time to between one half and one third that of a Corona-like MWSR using a token ring for arbitration.

Discussion: The authors of Pan et al. [2010] propose the FlexiShare network in order to address some of the short comings of the Firefly network. What is interesting about FlexiShare is that the number of data channels is decoupled from the number of nodes, which can lead to higher channel utilization.

4.2.5   DCAF

Topology: The Directly Connected Arbitration-Free (DCAF) design, described in Nitta [2011] and Nitta et al. [2012c], features waveguides that directly connect each source/destination pair of a 64-node network, creating a fully connected backbone; however, DCAF incorporates additional microring resonators in the transmitter section of each node, which are used to limit the number of destination nodes that a node can simultaneously be sent to. DCAF is in essence a many-to-one crossbar—a single node can simultaneously receive from multiple sources, but can send to only one. Figure 4.5(a) shows the equivalent network connectivity for a four-node DCAF.

Arbitration/Flow Control: Since the dedicated links make it possible for each node to receive messages from all other nodes simultaneously, no arbitration is required. DCAF essentially has a locally controlled demultiplexer in its transmit section, while Corona (as an example) has the equivalent of a receive multiplexer that must be globally arbitrated. Figure 4.5(b) shows how a 1:4 optical demultiplexer can be constructed using microring resonators. Figure 4.5(c) illustrates the DCAF transmitter section—in this figure λ_1 and λ_2 are being transmitted to node 2, while λ_3

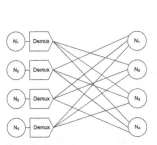

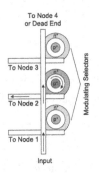

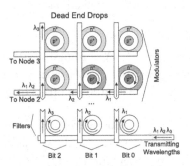

(a) Arbitration free crossbar (demux ensures that a node can send to only one other node at a given time.)

(b) 1:4 optical demux (signal is being output to Node 2.)

(c) TX section for node 4 (λ_1 and λ_2 are being transmitted to node 2, while λ_3 is not. A binary 011 is being transmitted to node 2.)

Figure 4.5: DCAF: 4-node network equivalent (a), 1:4 optical demultiplexer (b), and transmit section (c).

is not (in other words, node 4 is transmitting a binary 011 to node 2, using the modulating ones approach).

DCAF does not require arbitration in order to transmit a flit, and therefore it will not be subject to the limitations imposed by systems that require global clock synchronization. However, even though DCAF is arbitration-free, it does require flow control. This is accomplished using an ARQ scheme—if a flit arrives at a reception node and there is insufficient buffer space, the flit is dropped and the ACK is not sent back. A Go-Back-N (GBN) ARQ scheme was chosen over a conventional credit-based flow control approach since multiple flits can be in flight simultaneously on a single waveguide, and the round trip of a single link can be much greater than 2 cycles. The ARQ scheme allows for efficient flow control without the need for excessive buffering, and also provides a more reliable communication scheme, since lost or potentially corrupted flits can be retransmitted.

Buffering: Like Firefly, receive buffers are needed for each possible source, since multiple flits can be received simultaneously. Unlike Firefly, however, DCAF utilizes a shared FIFO that pulls from the per-source buffers, which reduces the amount of buffering neccessary for large node counts without sacrificing reliability (since the GBN ARQ is used for retransmission when there is insufficient buffer space and a flit is dropped).

Electrical/Optical Partitioning: The data transfer and control flow information (ACK) is all handled optically. The electrical control logic that moves the flits from the per-source receive buffers into the shared FIFO is also responsible for deciding which source to ACK when there are more than one outstanding flit to be acknowledged. Several configurations of receive buffer sizes and bus/crossbar conections between the receive buffers and share FIFO were analyzed in Nitta [2011].

Performance: According to Nitta et al. [2012c], DCAF can improve throughput on synthetic traffic patterns by ~60% when compared to a Corona-like crossbar using a Token Channel with Fast Forward arbitration protocol, and the peak energy efficiency of DCAF was six times higher.

Discussion: Considering the number of node connections (and hence the number of required waveguide crossings) and the assumed 0.1dB loss per intersection in Nitta et al. [2012c], a single-layer implementation of DCAF is not realizable (unless a very low-loss intersection could be constructed). However, it is possible to create directly connected networks like DCAF by using photonic vias and multiple photonic layers. Since the number of waveguides needed in DCAF grows quadratically with node count, in Nitta et al. [2012c] a more detailed evaluation of how DCAF might actually be laid out is presented. Included is an entire layout of a 16-node DCAF using a 16-bit bus, and a description of how a 64-node DCAF could be constructed from four groups of 16 nodes is also provided. In addition, in Nitta et al. [2012a] a generalization of DCAF called Directly Connected Optical Fabric (DCOF) is described, where it is possible to send to multiple destinations simultaneously.

In Joshi et al. [2009a], the authors propose a 64-tile photonic Clos network that, unlike many of the other proposed networks, was designed to be monolithically integrated with the processing cores. The findings presented in Joshi et al. [2009a] seem to agree with those in Nitta et al. [2012c], Nitta [2011], and Nitta et al. [2012a], which is that point-to-point channels are more energy efficient than global shared channels.

4.2.6   HYBRID PHOTONIC NOC

Topology: The basic building block of the hybrid photonic on-chip network proposed by Columbia University researchers [Shacham and Bergman, 2007; Shacham et al., 2007a,b, 2008] is a 2x2 photonic switch, referred to as a Photonic Switching Element (PSE), which is realized using a pair of active broadband microrings. In order to switch multiple-wavelength broadband signals, the rings are designed as comb-pass filters. A 4x4 switch is created using four PSEs and an electronic router (marked ER) as shown in Figure 4.6. A larger network is constructed by connecting these 4x4 switches using a 2D torus topology, as shown in the bottom of the figure. However, as is apparent in the figure, the 4x4 switch is not a non-blocking switch—a message going from South to East will block a connection from West to South and East to North. For full flexibility, a 2D mesh requires a 5x5 switch that is non-blocking.[3] In this work the designers deliberately choose to keep the photonic switch simple (instead of a full 5x5 crossbar), and compensate for the shortcomings by adding additional gateways and ejection and injection switches

---

[3]A non-blocking switch such as a crossbar allows all ports to be used simultaneously, while the fifth port provides connectivity to the local processing node.

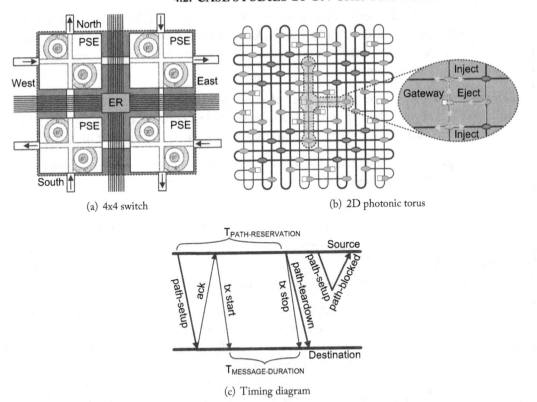

(a) 4x4 switch

(b) 2D photonic torus

(c) Timing diagram

Figure 4.6: A hybrid NoC. The basic building block is a 4x4 switch (a). A 2D folded torus topology (shown in dark ovals and thick lines), access points (shown as thin lines and light ovals), and gateways (shown as rectangles) is shown in (b). Timing diagram of the set up of a control path using electronics (c). (See Shacham et al. [2008] for original versions of figures.)

as shown in the bottom of Figure 4.6. This compromises flexibilty and adds some complexity to the routing algorithm, but keeps the photonics simple.

Arbitration/Flow Control: The electronic router is responsible for setting up a control path between the source and destination, controlling the corresponding PSEs, and for flow control. The principle of operation of the network is simple, and is captured by the timing diagram in Figure 4.6. A path-setup control packet is first sent from the source node to the destination using standard XY dimension-order routing, and once the path is set up, the destination node returns a small acknowledgement packet to the source node. The source node can start transmitting the data after the receipt of this acknowledgement packet. If the control packet used to set up the path is dropped by an intermediate router, a path block packet is sent back to the source, which frees up the resources and another attempt is made at a later time. Note that the proposed path setup mechanism using electronic routers can also handle flow control—basically the destination node can withhold the acknowledgement packet if it cannot accept new messages. Once the transmission is complete, a small control packet is sent to the destination in order to tear down the path and free up the resources.

Buffering: The data packets are buffered at the source node electronically until a path to the destination is available. The buffering needs in the electrical network are minimal since it is only used as a control plane, and thus only has to store control packets (which are much smaller).

Electrical/Optical Partitioning: As stated above, an electronic control plane is used to set up a path between the source and destination nodes, and a photonic dataplane is used to send the actual data.

Performance: The researchers simulated a 36-core Chip MultiProcessor (CMP) in a 22nm technology node, assuming a chip with 2cm on the side and 3.3mm x 3.3mm cores, with 36 gateway access points. They found that this approach clearly shows the benefit of photonic networks in terms of bandwidth and latency, especially on large messages.

Discussion: The main advantage of the proposed hybrid network is its simplicity, as it leverages existing electronic on-chip network technology to do the heavy lifting (arbitration, flow control, path setup, etc.) and uses photonics to move data rapidly, once the path is available (about 15.4ps per mm latency). Also, data buffers are not required at the intermediate nodes since circuit switching is used for data transport. However, setting up the control path can take several nanoseconds, especially under high load. Thus, this network is best suited for large granularity data transfers (such as Direct Memory Access (DMA)) and does not perform as well on applications that require short messages (like cache-coherency traffic). The optimal message size depends on many factors, such as the size of the network, the latency of the individual components, and the bandwidth of the injection/ejection ports. Further details on the impact of message size, power estimation, and performance on scientific workloads can be found in Shacham et al. [2008] and Hendry et al. [2009].

CHAPTER 5

Challenges

There are two key technical challenges facing photonic on-chip interconnects—getting the microrings tuned to the desired resonance frequency, and then keeping them there. Microrings must be fabricated with great precision in order to resonate at exactly the necessary frequency, and they are also very sensitive to variations in temperature. This is particularly challenging if a computing system design uses hundreds of thousands of these components, and they are integrated into (or in close proximity to) a package which contains the heat-generating processors themselves.

At a high level there are two (non-exclusive) ways to deal with this problem—either keep the rings "in tune" (resonating correctly), or else leverage the unique properties of photonics to provide sufficient error detection/correction such that rings that are no longer in tune won't cause incorrect behavior. In this chapter we will present an overview of the technical challenges themselves, as well as various approaches proposed by the research community to deal with the problems.

5.1 PROCESS VARIATIONS

The high refractive index of silicon, while useful for concentrating and confining light, leads to devices that are very sensitive to process variations in fabrication. This sensitivity makes it difficult to achieve uniformity from chip to chip, from device to device, and even within a single device [Selvaraja, 2011]. Non-uniformities in the critical dimensions of fabricated microrings can cause them to resonate at a different wavelength than desired, since the resonance wavelength is dependent upon the microring radius. In addition to variations in the microring geometries, roughness in the waveguide top surface and sidewalls can cause increased path attenuation due to light scattering [Lipson, 2005; Selvaraja, 2011]. Passive or post-fabrication techniques such as using Ultraviolet (UV) light to correct for fabrication errors have been proposed [Kokubun et al., 2010; Schrauwen et al., 2008; Zhou et al., 2009b], and these approaches have distinct advantages—for example, once all the microrings are adjusted, only the sensitivity to temperature needs to be addressed. However, this approach requires each ring to be analyzed and adjusted individually, so it is not clear how practical this technique will be when scaled to the system level.

IN DEPTH: PROCESS VARIATION DRIFT & LOSS DETAILS

The spectral response for a silicon optical device depends upon the effective index of refraction n_{eff}. Any change in the height or width of the waveguide shifts the effective index of refraction, and thus will change the spectral response. The spectral drift can be estimated as $\frac{\Delta\lambda}{\lambda} \approx \frac{\Delta n_{eff}}{n_{eff}}$, where the change in effective index of refraction per change in waveguide width $(\frac{\Delta n_{eff}}{\Delta W})$ or waveguide height $(\frac{\Delta n_{eff}}{\Delta h})$ must be determined through simulation [Selvaraja, 2011].

According to Lipson [2005], the propagation losses α can be modeled as Rayleigh scattering in fibers: $\alpha = \frac{\sigma^2 k_0^2 n_{core}}{n_{eff}} \cdot \frac{E_S^2}{\int E^2 dx} \cdot \Delta n^2$, where σ is the size of the roughness, k_0 is the free-space wavenumber, E_S is the field intensity at the core/cladding interface, E is the field intensity at a position x along the cross section of the waveguide, n_{eff} is the effective index of refraction, and Δn is the index contrast between the core and the cladding.

5.2   THERMAL ISSUES

The wavelength that an individual microring responds to is set during fabrication and is dependent upon the index of refraction of the microring—however, the refractive index of silicon is very sensitive to temperature. Changes in the microring index of refraction cause the resonance wavelength to drift spectrally approximately 0.09nm/°C. To put this into perspective, a 10Ghz data channel centered at 1550nm occupies ~0.08nm of the spectrum, so a 1°C change will shift the microrings resonance completely off its desired channel. Clearly the thermal sensitivity of microrings is a challenge that must be overcome, since keeping the microrings bound within a tight temperature window is not practical.

IN DEPTH: REDUCED THERMAL SENSITIVITY

Researchers have been making progress on ways to reduce the thermal sensitivity of microring resonators. In Zhou et al. [2009b], the authors demonstrated that by using Polymethyl Methacrylate (PMMA) as an upper cladding on microring resonators, they were able to reduce the temperature sensitivity to 27pm/°C, or to about a quarter of that of the baseline microring. In Teng et al. [2009], it was experimentally demonstrated that overlaying a polymer layer around a 15μm radii microring reduced the thermal sensitivity to less than 5pm/°C. The authors of Raghunathan et al. [2010] developed the principles for an athermalized microring, and fabricated a 20μ radii microring with a polymer cladding that had a thermal sensitivity of 0.5pm/°C. Finally, Guha et al. [2010] demonstrated a microring with a 40μm radius which was coupled to a 796μm x 230μm Mach-Zehnder Interferometer (MZI) and had an operational temperature range of 80°C. (In other words, this design essentially showed "temperature insensitivity" over a range of 80 degrees, meaning it was athermal in that range.)

The refractive index (n) of silicon changes due to changes in ambient temperature (ΔT), which can be modeled as $\Delta n \approx 1.84 \times 10^{-4} \times \Delta T$. This positive Thermo-Optic (TO) coefficient of silicon is counteracted in the works of Zhou et al. [2009b], Teng et al. [2009], and Raghunathan et al. [2010] by using a cladding that has a negative TO coefficient. Specifically, they use PMMA, a PSQ-LH polymer, and a hyperlinked fluoropolymer with TO coefficients of -1.2×10^{-4}, -2.4×10^{-4}, and -2.65×10^{-4}, respectively.

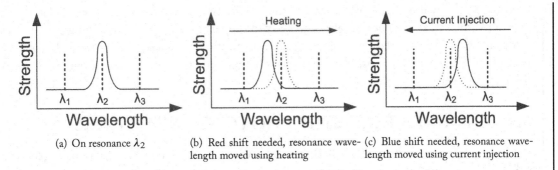

(a) On resonance λ_2

(b) Red shift needed, resonance wavelength moved using heating

(c) Blue shift needed, resonance wavelength moved using current injection

Figure 5.1: Microring resonance vs. wavelength – (a) shows on resonance λ_2, (b) shows that red shift is needed (solid line) and heating is applied (dotted line), and (c) shows blue shift is needed (solid line) and current injection is used (dotted line).

5.3   TRIMMING

A technique known as *trimming* can be used to dynamically move the resonance frequency of a microring toward either the red or the blue, which allows both thermal drift and errors due to proccess variations to be corrected. Moving the resonance frequency toward the red can be accomplished by heating the microrings [Ahn et al., 2009], a technique that exploits the index of refraction change in silicon described previously. Heating is done using thin-film platinum surface heaters near waveguide sections [Frey et al., 2006]. The resonance can be shifted toward the blue by increasing the current in the n^+ region of an active modulator (this is the same technique used as a modulation mechanism, described in Section 2.1.3). Figure 5.1 illustrates how resonance frequency changes when heating (Figure 5.1(b)) and when injecting current (Figure 5.1(c)). Unfortunately, both these trimming techniques can result in a dramatic increase in the overall power requirements of a network—for example, in Ahn et al. [2009] they estimate that a total of approximately 26W is necessary for trimming of the Corona network (which is about 54% of the estimated 48W total network power.)

In much of the on-chip photonics interconnect literature (e.g., Ahn et al. [2009]; Joshi et al. [2009a]; Pan et al. [2010]), the power required to perform trimming has been treated primarily as a fixed cost per microring. This fixed cost was then multiplied by the number of microrings in the design in order to arrive at a global estimate. However, in Nitta et al. [2011a] it is shown that the power required for trimming is *not* a per/ring fixed cost—the energy required to shift the resonance to the red via heating has a non-linear relationship with the number of rings, and shifting the resonance to the blue using current injection heats the rings and can quickly lead to thermal runaway. This investigation revealed that when using heating, the die size is more important than the actual number of microrings (primarily due to the thermal conductance of the entire die), and that the feedback mechanism caused by current injection can be dramatically reduced if the microring thermal sensitivity is decreased. Regardless of the trimming mechanism used, a full thermal analysis of the entire photonic network should be completed in order to determine the actual amount of power necessary to support trimming; the trimming power could not only be underestimated (as in case of current injection feedback), but potentially could be overestimated as well if it is assumed heating is being used and the cores are providing some of the required heat.

5.4   RESILIENT ON-CHIP PHOTONIC NETWORKS

In many areas, such as hard disks, flash memory, wireless communication, and small geometry DRAMs, it is common to try to create a reliable communication link using underlying components that may be unreliable, instead of attempting to make each individual component reliable. For reasons explained in the previous sections, it may not be practical to analyze and adjust every microring using UV light to correct for fabrication errors, and the amount of power at the system level needed to perform trimming may be significant. Thus, in Nitta et al. [2011b] ways to make optical links *resilient* were explored—in other words, approaches that would make it possible to tolerate some malfunctioning rings and still be able to communicate reliably using an on-chip optical link. This approach is particularly appealing because of the surplus bandwidth that exists in a WDM-based optical network, which can potentially be leveraged to improve the resilience of the network. In order to accomplish this, a better understanding of the nature of faults in the photonic realm is needed.

5.4.1   PHOTONIC LINK FAULT MODELS

The faults that occur in photonic components can lead to a variety of different bit errors. In this section the various low-level faults due to problems with the microring resonators will be abstracted out into a set of "fault models" that can be used by the architect when implementing a resilient photonic network.[1]

Microrings that do not resonate at their designed spectral position and waveguides with increased attenuation will be abstracted further. Microrings that do not resonate as designed will be considered faulty, which will happen if they are resonating to the wrong wavelength, the signal attenuation is too great, or both. The two exclusive cases are illustrated in the graphs in Figure 5.2. In this figure the dashed lines show the desired function, while the solid lines show the actual. Figure 5.2(a) shows that when the microring is not accurately tuned to the desired wavelength λ_2, the amount of λ_2 that appears on the *drop* port is very small—only the amount of the λ_2 line that lies below the solid black line. This misalignment could be the result of thermal drift, process variations during fabrication, insufficient trimming, etc. Figure 5.2(b) shows the power to the *drop* port of a microring that is excessively attenuating the signal, which results from improper fabrication or possibly too much current injection.

---

[1]We will follow the terminology and flow presented in Parhami [1994]: Defect → Fault → Error → Malfunction → Degradation → Failure.

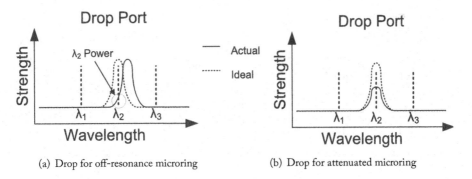

(a) Drop for off-resonance microring

(b) Drop for attenuated microring

Figure 5.2: Drop for off-resonance (a) and attenuated (b) microrings designed to resonate on λ_2.

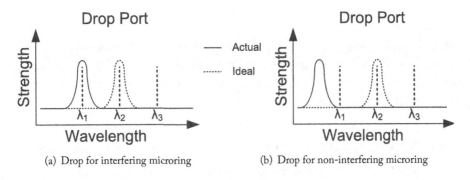

(a) Drop for interfering microring

(b) Drop for non-interfering microring

Figure 5.3: Drop for interfering (a) and non-interfering (b) microrings designed to resonate on λ_2.

Some of the faulty optical components in an on-chip network could potentially be made to work correctly by increasing the amount of system power used. Off-resonance microrings that are being trimmed, for example, simply require more signal power to operate correctly. Waveguides that have greater path attenuation may also be compensated for by increasing the photonic power, enough so that sufficient power reaches the photodetectors. In this model, we focus on the cases for which addressing the fault will not be as simple as increasing the power.

Microrings that do not resonate at their designed spectral position (as in Figure 5.2(a)) can be put into one of two categories: *interfering* or *non-interfering*. Interfering microrings are those whose frequency has drifted so far that they are actually resonating at another wavelength channel. Figure 5.3(a) shows the power to the *drop* port of an interfering microring that is designed to resonate at λ_2, but is interfering with λ_1. Non-interfering microrings are those that do not resonate at the desired wavelength, but do not interfere with any other wavelength either. Figure 5.3(b) shows the power to the *drop* port of a non-interfering microring that is designed to resonate at λ_2, but is resonating below λ_1. Microrings that have increased attenuation are considered to be non-interfering microrings, since the end result is the same as a slightly off-resonance

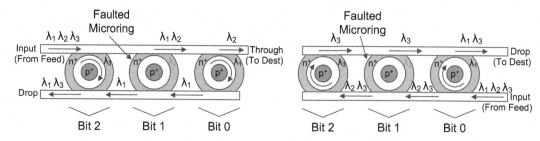

(a) Modulating zero (removing wavelengths). Attempting to send 000, actually sending 100

(b) Modulating ones (adding wavelengths). Attempting to send 111, actually sending 101

Figure 5.4: Single errors due to faulty microring for modulation of zeros (a), and modulation of ones (b).

non-interfering microring—in both cases, a diminished amount of the desired wavelength appears on the output port. Microrings that are partially interfering could also be viewed as being two non-interfering microrings, since both cases result in neither of the two wavelengths being properly transmitted or received.

5.4.2 LINK COMPONENT STRUCTURE-DEPENDENT ERRORS

The types of errors that will result from faults depends upon the structure of the link components. Here the focus will be on the transmitter and receiver sections of the on-chip optical networks, since (as we discussed in Chapter 2) the proposed networks all have similar transmitter/receiver structures and differ primarily in the interconnection topology. As we discussed in Section 2.1.3, transmitting data is done in one of two ways: by actively modulating ones (transitioning wavelengths from the *input* waveguide to *drop* waveguide), or by actively modulating zeros (removing wavelengths from the *through* waveguide). The receiver section for a link will consist of a set of microring resonators that are either always on-resonance (as in Vantrease et al. [2008]), or enabled whenever a message is sent (as in the reservation-assisted SWMR, proposed in Pan et al. [2009]).

Non-Interfering Microring Fault Errors

Non-Interfering faults do not move the desired wavelength from the *input* port to the *drop* port (or *add* port to *through*), but do not transition any other wavelength either. Thus, a non-interfering faulty microring that is in the receiver section will result in zeros always being received for that bit, since the proper wavelength will never transition from the *input* port to the *drop* port. This is essentially a "stuck-at-zero" fault, and only results in a bit error when a one is being sent on that bit.

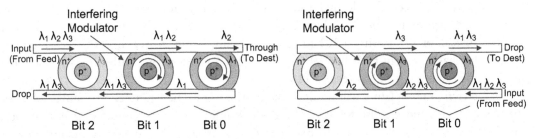

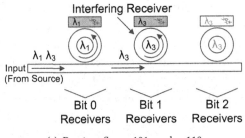

(a) Modulating zeros. Attempting to send 100, actually sending 101

(b) Modulating ones. Attempting to send 011, actually sending 101

(c) Receiver. Sent a 101, reads a 110

Figure 5.5: Double-bit error from interfering microring fault for modulation of zeros (a), modulation of ones (b), and reception (c).

The types of errors generated by a non-interfering faulty microring in the transmitter section will depend upon the method of modulation. In the case where zeros are actively modulated, a faulty microring will result in the wavelength always being present at the destination (a one will always be detected, which corresponds to a "stuck-at-one" fault). This is shown in Figure 5.4(a), where a three-bit transmit section is attempting to send all zeros. The Bit 1 modulator is faulty, so λ_2 is not being removed from the waveguide.

In the case where ones are actively modulated, a faulty microring will result in its resonant wavelength never being present at the destination (a zero will always be detected, which corresponds to a "stuck-at-zero" fault). Figure 5.4(b) illustrates a three-bit transmit section that is attempting to send all ones—again, Bit 1 has a faulty modulator and therefore is not transitioning λ_2 from the *input* feed to the *drop*.

Interfering Microring Fault Errors
Interfering faults are much more problematic than non-interfering faults. It is possible for double-bit errors to occur when an interfering faulty microring is involved, for example. Figure 5.5 illustrates double errors for both forms of modulation and for reception. Figure 5.5(a) shows a three-bit transmit section attempting to transmit the value 100, but Bit 1 is interfering with Bit 2

(λ_3 is removed instead of λ_2), leading to the value 010 being sent. In Figure 5.5(b), the transmit section is attempting to send the value 011, but again Bit 1 is interfering with Bit 2 (λ_3 is transitioned instead of λ_2) causing a 101 to be sent. Finally, Figure 5.5(c) shows a three-bit receive section that has been sent the value 101, but since Bit 1 is interfering with Bit 2 (λ_3 is removed by Bit 1 instead of by Bit 2), it reads a 110.

Interfering modulators will result in the interfer**ing** bit being "stuck-at"(similar to a non-interfering fault), and the interfer**ed** bit being a logical function of the interfering and interfered bits (similar to a *bridged* fault). In the case where zeros are actively modulated, the interfered bit will be a logical AND of the interfering and interfered bits, since either modulator will remove the wavelength in the case of a zero, and only both bits being a one will result in the wavelength passing unperturbed. In the case where ones are actively modulated, the interfered bit will be a logical OR of the interfering and interfered bits. The case where ones are actively modulated is symmetric to that of the case where zeros are actively modulated, as one might expect.

In the receive section, microrings that are resonating at another wavelength may or may not actually be interfering. Figure 5.5(c) shows that Bit 2 cannot interfere with Bit 0, even if it is resonating at λ_1. A microring resonating at another wavelength, but not interfering, behaves like a non-interfering microring ("stuck-at-zero"). However, in the case where one microring *is* interfering with another, the interfered bit will manifest as a "stuck-at-zero," and the interfering bit will receive the interfered bit's information.

5.4.3   UNIDIRECTIONAL BIT ERRORS

The choice of modulation and reception topology can lead to an asymmetry of errors when certain faults occur. It is clear that interfering faults can lead to double-bit errors, but non-interfering faults lead to errors in a single direction. Non-interfering faults for receivers and modulators that actively modulate ones will only cause $1 \rightarrow 0$ bit errors, since they are "stuck-at-zero" faults. Increased path attenuation can also lead to $1 \rightarrow 0$ bit errors (since insufficient photonic power to switch from $0 \rightarrow 1$ reaches the photodetector). On the other hand, modulators that actively modulate zeros will have faults that yield $0 \rightarrow 1$ bit errors. Non-interfering faults in components result in the following unidirectional bit errors:

Modulator (Active Zeros): Light will not be successfully removed from the *through* waveguide. When zeros are sent, a one will be received ($0 \rightarrow 1$ bit error).

Modulator (Active Ones): Light will not be successfully transitioned to the *drop* waveguide. When ones are sent, a zero will be received ($1 \rightarrow 0$ bit error).

Receiver: Light will not be successfully transitioned from the *input* to the photodetector. When ones are sent to it, a zero will be received ($1 \rightarrow 0$ bit error).

Waveguide: Increased waveguide attenuation results in insufficient light being received at the end of the waveguide. When ones are sent, a zero will be received ($1 \rightarrow 0$ bit error).

The type of single-bit errors that will occur in a photonic link can be designed to be unidirectional if the correct link component structure is chosen. This is important because unidirectional errors can be dealt with more efficiently, and if the architect is willing to give up some bandwidth and separate the channels more, it may be possible to minimize/eliminate interfering faults.

5.4.4   MEAN TIME BETWEEN FAILURES

Improving communication link reliability can be accomplished by increasing the probability that each transmission will be received correctly, by retransmitting until the transmission is received correctly, or both. Increasing the probability of a correct reception can be done using fairly straight-forward techniques, such as reducing the error rate (reducing the device fault rate) and/or adding bits in order to correct for errors. Retransmitting messages until they are properly received is a little more complicated, since it requires a feedback communication link and a communication protocol.

There are certain unique properties of photonics that can impact the choice of error correction schemes. For example, since much of the power consumed in a photonic link is static (the external laser and microring trimming, for instance), the relative power penalty of retransmission or of error detection/correction techniques may be much lower than it would be with electrical links. In photonic systems, the high static overhead means the cost of transmitting data is mostly pre-paid—the more you transmit, the lower the average cost/bit becomes.

In Nitta [2011], a range of approaches to providing reliable links were studied. Common methods of providing reliable data transmission over an unreliable communication channel, such as the use of an ARQ protocol, were considered—however, an ARQ protocol alone will not guarantee reliable communication for all fault sources. If the faults are permanent, for example, ARQ protocols will unsuccessfully repeat transmissions until the maximum retransmission count is reached. To circumvent this problem, a Hybrid Automatic Repeat reQuest (HARQ) protocol can be employed, which utilizes Forward Error Correction (FEC) in order to correct a small number of errors and only requests retransmission for uncorrectable cases. Thus, a HARQ protocol can make on-chip networks reliable even in the presence of some permanent faults, as long as they are correctable by the FEC.

Error Correction Codes

Error correction codes such as Cyclic Redundancy Check (CRC) were shown to be poorly suited for the photonic environment in Nitta [2011], since communication is not a serial stream of bits (making burst errors less likely), and the block length is relatively short. Furthermore a CRC code is typically calculated in hardware using a Linear Feedback Shift Register (LFSR), which would have difficulty keeping up with the communication rates of on-chip networks (although parallel implementations of LFSRs do exist [Campobello et al., 2003; Lu and Wong, 2003]).

However, Berger codes [Berger, 1961] can detect any number of unidirectional bit errors with the addition of $k = \lceil log_2(n + 1) \rceil$ check bits, where n is the number of data bits. The efficiency of the coding makes Berger codes good candidates for use in this setting—unfortunately, Berger codes require the computation of the *weight* of the codeword, which is very expensive.

Extended Hamming codes for Single Error Correction and Double Error Detection (SECDED) have been used in a number of memory systems, including the CRAY-1. The same SECDED codes can be utilized as a Triple Error Detection (TED) code, if no correction is performed. A SECDED or TED code can be implemented for 32 and 64 data bits with the addition of 7 and 8 check bits, respectively.

Another approach to error detection is to use multiple signals to transmit a single bit of information. This approach is commonly used in high-speed communications, such as Low Voltage Differential Signaling (LVDS). This is essentially an N choose K (NcK) encoding—for example, LVDS is a 2c1 encoding, since only one of the two signals will be a one at any given time. NcK encodings can detect all odd number of bit errors, and may be able to detect some even number of bit errors. Moreover, NcK encoding can detect *any* number of unidirectional errors.

As discussed previously, HARQ requires FEC. One possible FEC code that could be used in HARQ is the extended Hamming SECDED code—another approach is to combine the NcK encoding with either a parity block or a Reed Solomon code. Since any odd number of bit errors can be detected with an NcK encoding, the detected errors could be treated as block erasures, and an additional parity block could be used to recover from a single erasure (as is commonly done in Redundant Array of Independent Disks (RAID)-5).

IN DEPTH: REED SOLOMON

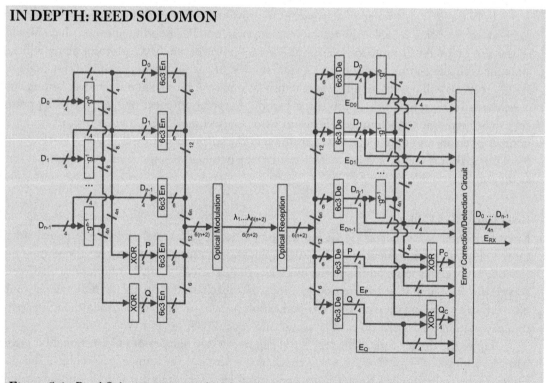

Figure 5.6: Reed Solomon integrated with 6c3 encoding circuit.

The redundancy could be extended to protect against double erasures as in RAID-6, as long as the size of the Galois Field (GF) being used for the Reed Solomon code blocks is large enough. A **GF**(2^n) can cover $2^n - 1$ data blocks; therefore, a 2c1 code could only cover a single data block, while a 6c3 could cover 15 data blocks $(2^4 - 1)$ or up to 60 bits of data. Reed-Solomon codes can detect and even correct for block corruptions. The equations $P = D_0 + D_1 + ... + D_{n-1}$ and $Q = g^0 \cdot D_0 + g^1 \cdot D_1 + ... + g^{n-1} \cdot D_{n-1}$ are used to calculate the Reed-Solomon code.

In these equations, "addition" is handled by an XOR, and "multiplication" is done in the Galois Field. At first glance it may seem that calculating the GF multiplication may be too complex, but since we are proposing only a 4-bit code word and the multiplication is being done with a constant value, it can be realized with a simple look-up-table. Figure 5.6 illustrates a potential Reed-Solomon circuit that utilizes 6c3 block encoding. Details of the error correction/detection circuit have been omitted for brevity, but would primarily be composed of a network of multiplexers.

Mean Time Between Failure Analysis

In order to justify choosing one encoding scheme over another, one must know both the microring fault rate and the rate of interfering vs. non-interfering faults. Since this information is not yet available, one can take a different approach; one can determine the fault rate that microrings must attain in order to meet a particular Mean Time Between Failure (MTBF) for a single link, and also for an entire network (such as a photonic torus). These calculations can not only guide architects in the choice of encoding schemes once microring resonators mature, but these results also provide goals and targets for device researchers and manufacturers.

The MTBF for a link can be calculated given the fault rate, the probability that a fault is interfering/non-interfering, and the average rate of error for various proposed detection/correction schemes. In Nitta [2011], a detailed simulator is described that is able to pro-

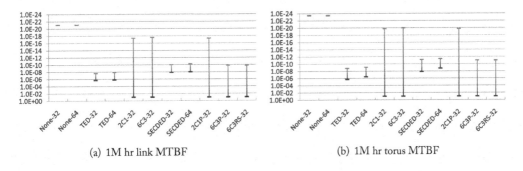

(a) 1M hr link MTBF   (b) 1M hr torus MTBF

Figure 5.7: Required microring fault rate to attain 1M hr MTBF for a link (a) and a 8-ary 2-cube torus (b) by encoding scheme.

vide this information. Since the target MTBF is known, the fault rate is simply varied until the target MTBF is achieved. Figure 5.7 shows the required microring fault rate given a particular encoding and a desired MTBF of 1M hours, for a single link (5.7(a)) and an 8-ary 2-cube Torus (5.7(b)). The torus was assumed to have direct links between the nodes (no microrings were assumed for routing). The Y-axis is the required fault rate that must be attained, with a lower fault rate (requiring higher-quality microrings) being higher on the axis.

The spread of the values is due to different rates of interfering faults. The values at the lower fault rates assume the probability of a fault being interfering has a uniformly random distribution—in other words, the probability of an interfering fault is the percent of the FSR that other channels occupy. The values requiring a higher fault rate, on the other hand, assume the resonance point of a microring will drift from the desired point based on a normal distribution, centered at the desired resonance frequency (which yields a dramatically lower rate of interfering faults.)

The results show that the NcK encodings such as the 2c1-32 or the 6c3rs-32 are the best choice when fault rates are very high, but the rate of interfering faults is very low. The Hamming codes are best when the fault rates are moderately high, with TED winning out over SECDED if correcting for fabrication errors is not a concern. In order for nanophotonic links/topologies to meet a 1M hour MTBF without using error detection or correction schemes, microrings will need to be fabricated such that fault rates are in the range of 10^{-21}–10^{-24}/cycle. The conservative assumption used in the simulations (that any undetected bit error will result in a failure) means these numbers are probably a little high, but it is unlikely that the actual bit error rate that results in failure will change these results by very much—certainly not by orders of magnitude. Given the current immature state of the technology, it is clear that some type of error detection scheme will be needed if large-scale microring resonator-based networks are to become a reality. Microring-based photonic networks that do not implement error detection or correction schemes will be inherently unreliable due to their low MTBF.

CHAPTER 6

Other Developments

The focus of this book is on on-chip photonic interconnects, but there are many exciting developments in related areas that should be of interest to a computer architect. Some of these developments, such as those in implementation technologies and devices (particularly lasers), could impact the design of future on-chip networks. Others have the potential to overcome various bottlenecks in computer system design. In this chapter we will briefly describe some of these developments.

6.1 ON-CHIP NETWORK DEVELOPMENTS

Research in the following areas may have a direct impact on the success of on-chip networking.

6.1.1 MONOLITHIC CMOS INTEGRATION

The overwhelming majority of semiconductors (92% of 300mm wafers) use bulk-CMOS, which continues to improve with advances in lithography and fabrication processes. Unfortunately, no front-end photonic integration solution has yet been proposed for bulk-CMOS processes. Thus, there are two choices when implementing photonic integrated circuits today—use a SOI-CMOS process with a thick Buried Oxide (BOX) layer (an order of magnitude thicker than in standard SOI-CMOS), or use a separate layer of silicon nitride deposited on top of the metal stack [Levy et al., 2009; Shaw et al., 2005]. These specialized silicon photonic dies can then be attached to the rest of the CMOS electronics using wafer bonding, or perhaps 3D integration technology like TSVs. Clearly, there is an advantage to implementing photonics circuitry on a separate die—ad-hoc fabrication processes could be used to implement ridge waveguides or germanium photodetectors, for example, which are not possible to do with standard CMOS, SOI or bulk processes. However, there are disadvantages as well. The thick BOX layer degrades the performance of transistors, and its thermal impedance reduces the amount of heat that can be dissipated, which in turn limits the thermal budget of the electronics. More importantly, the electronics required to drive the photonic interconnects cannot take advantage of any improvements in CMOS technology. The lack of an integrated CAD flow and manufacturing workflow also increases cost and complexity.

Researchers at MIT have begun to address these challenges by developing a monolithic photonic integration process, where photonic devices are implemented using standard CMOS logic and DRAM processes [Orcutt and Ram, 2010; Orcutt et al., 2011]. They have demonstrated photonic devices with 3dB/cm waveguide loss using an existing commercial 45nm SOI-CMOS process. The photonic devices are monolithically integrated with electronics in the same physical device using a standard CAD flow [Orcutt et al., 2012].

6.1.2 LASERS

Despite their widespread use in telecommunications, consumer, and industrial applications, lasers are a relatively recent invention, with semiconductor lasers first demonstrated in 1962. Consequently, there are many hurdles when it comes to the use of lasers in on-chip networks. First, the electrical efficiency of lasers is poor—around 10% to 30% at most, with commercial wavelength-stabilized lasers from the telecom world exhibiting a pathetic 1% wall-plug efficiency. Second, since silicon is an indirect band gap material, it does not lend itself to *lasing* easily, which makes the creation of on-chip lasers difficult. Third, generating multiple wavelengths of light for use in DWDM is a challenge, especially at high data rates and with acceptable wall-plug efficiency and without crosstalk. Active research is underway to overcome these challenges, and we will briefly summarize some of the numerous advances that have been reported in recent years. The interested reader should see Beausoleil [2011] for more details.

One approach to building a silicon laser is to take advantage of stimulated Raman scattering, as demonstrated by Jalali and Fathpour [2006] and Boyraz and Jalali [2004]. However, the efficiency of such silicon Raman lasers is still quite low and needs to be improved for it to be practical. In Liu et al. [2010b], it was shown that Germanium (an indirect band gap material) can be made into a pseudo-direct band gap material by using tensile strain and n-type doping, and demonstrated that it is possible to construct a room temperature, optically pumped Ge-on-Si laser (laser pumping is the transferring of energy from an external source into the gain medium of the laser; the external pump energy is usually provided as pulses of light or electrical current). Building upon the work of Liu et al. [2010b], MIT researchers [Camacho-Aguilera et al., 2012] showed that Ge-on-Si lasers could be electrically instead of optically pumped; this is an important experimental result since it demonstrates that an additional light source will not be necessary to create a Ge-on-Si laser. This electrically pumped Ge-on-Si multimode laser had 1mW output at room temperature, which is a very encouraging development toward an on-chip laser that could be used with photonic interconnects. Another possibility is the heterogeneous integration (via bonding) of a direct bandgap material such as Indium-Phosphide (InP) onto a SOI-substrate [Van Campenhout et al., 2007, 2008], or bonding individual III-V laser chips onto silicon [Liang and Bowers, 2010]. Recent developments in transfer-printing epitaxial layers of compound semiconductors makes this more practical [Kelsall, 2012].

There are also interesting developments in generating a frequency comb for WDM-based photonic interconnects, especially for on-chip applications, that merit discussion. Quantum dot Fabry-Perot lasers are capable of producing a large number of discrete wavelength channels, separated from their nearest neighbors by the FSR frequency [Gubenko et al., 2007; Wojcik et al., 2009]. They use a single cavity to generate multiple laser channels simultaneously, so they are likely to be more efficient as well, since only one cavity has to be stabilized. Other work from Cornell [Foster et al., 2011] shows how to make broad-bandwidth optical frequency combs from a CMOS-compatible integrated microresonator, using parameteric oscillators to generate multiple wavelengths. Both the microresonator and the coupling waveguide are fabricated monolithically in a single silicon nitride layer using electron-beam lithography. A fixed comb of frequencies to support WDM can also be generated by externally modulating a conventional laser, as described in Zhou et al. [2009a].

6.1.3   PLASMONICS

Surface plasmons are coherent electron oscillations (or quasi-particles formed from the coupling of a photon and a electron wave) at the interface of a metal and a dielectric. Plasmons maintain many of the benefits of optical interconnects (such as bandwidth and energy), but have the additional benefit of not requiring waveguides with a width that must be on the scale of the wavelength of light [Dionne et al., 2010; Wassel et al., 2012]. This means that they can propagate in smaller areas, and can have rapid changes in the waveguide geometry without significant energy loss; however, these benefits come at a cost of significantly higher path attenuation.

Plasmonics, like grating couplers, may be suitable for use as a wavelength-blind photonic via. Plasmonics have the capability to drastically change the direction of light, which could be useful when changing layers; unfortunately, plasmonics suffer from high path attenuation. This attenuation is derived from the propagation length L_{SP}; the propagation length is defined as the length in which the Plasmon intensity has decayed to $\frac{1}{e}$ of the original strength. $L_{SP} = \frac{1}{k_x''}$, where $k_x = k_x' + i k_x''$ and k_x is the plasmon wave vector (surface plasmons are solutions to Maxwell's equations in planar geometries). An attenuation of $\sim$0.2dB/μm (equivalent to $\sim$20μm propagation length) is often assumed for plasmonics [Dionne et al., 2010], and over the relatively short distances required for an inter-layer via (assumed less than 10μm) the loss experienced by a plasmonic-based photonic via may be acceptable.

Energy efficient plasmonic modulating structures, such as the Compact Modulator [Cai et al., 2009] and the PlasMOStor [Dionne et al., 2009], have been proposed and, in the case of the PlasMOStor, actually fabricated. While these modulators require only a few fJ/b to modulate, they are wavelength-blind and therefore not suitable on their own for WDM; however, there has been interesting work done recently [Wassel et al., 2010, 2012] that uses passive microring resonators in combination with these plasmonic modulators to realize a WDM-compatible modulator. The approach proposed in Wassel et al. [2010] and Wassel et al. [2012] uses cascaded passive microrings to filter and to reinsert the individual wavelengths, and utilizes the plasmonic modulators to modulate the individual wavelenths. Figure 6.1(a) illustrates the proposed hybrid plasmonic modulator. This approach has the advantage of lower modulating power, since the capacitive load of the plasmonics is much smaller than the microring resonator. In addition, if the modulators are constructed as shown in Figure 6.1(b), this could allow the modulation circuitry to be placed closer to the rest of the electronics—the limit would become the waveguide pitch instead of the pitch of microring resonators (which would reduce the transport power discussed in

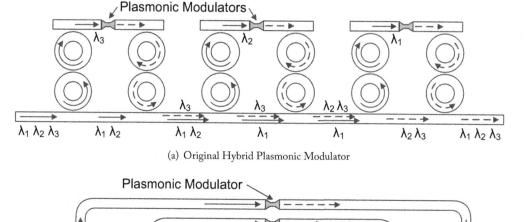

(a) Original Hybrid Plasmonic Modulator

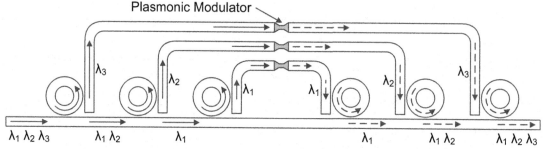

(b) Minimal Transport Hybrid Plasmonic Modulator

Figure 6.1: Hybrid plasmonic microring modulator for WDM. The original layout proposed in Wassel et al. [2012] shown in (a). Our proposed layout modification to minimize transport power is shown in (b). The individual wavelengths are filtered on the left by the set of passive microring resonators. The individual wavelengths are modulated using plasmonics and then recombined on the single waveguide using a second set of passive microring resonators.

Chapter 3). While the hybrid plasmonic approach reduces the modulation power, it does increase the photonic path attenuation due to the increased number of on- and off-resonance microrings, as well as the additional loss incurred by the plasmonic devices. This increased path attenuation may be mitigated by using fewer wavelengths modulated at higher frequencies, which may be beneficial as the results in Chapter 3 indicate.

Structures such as surface plasmon nano-antennas [Cao et al., 2010; Tang and Miller, 2009; Yousefi and Foster, 2012] have the potential to dramatically improve the energy efficiency of receiving photonic information. The capacitances of these nano-antennas can be on the order of a few attofarads (that is about a thousand times lower than the assumed few femtofarad capacitances that the PIN photodiodes/phototransistors may approach). Many of the nano-antennas demonstrated thus far have been for a single wavelength and narrowband [Maksymov et al., 2012]; this means that it might be possible for nano-antennas to be designed to be wavelength selective, removing the need for microring resonators and photodetectors in the receiver altogether.

6.2   SYSTEM-LEVEL DEVELOPMENTS

In this section we will describe some interesting developments that explore the use of photonic interconnects to address some well-known bottlenecks in computer systems.

6.2.1   OFF-CHIP I/O

Off-chip I/O bandwidth is the product of the number of pins on the chip and the data rate per pin. The ITRS 2012 roadmap (see Table ORTC-4 Performance and Packaged Chip Trends in Semiconductor Industry Association [2012]) indicates that the maximum number of "pins" on a chip is not changing, even in the long term—it will be around 3072. Of these, approximately two thirds are reserved for power and ground, leaving only about 1,000 or so pads for signal I/O. Furthermore, driving a signal off-chip requires driving a significantly higher capaticance (which grows linearly with distance), inherently limiting the data rate. Consequently, the off-chip I/O bandwidth is a serious impediment going forward, particularly since technology scaling is making it possible to place more and more processing cores on a die.

Fortunately, photonic interconnects have the potential to address both of these constraints. The pin limitation can be overcome by using DWDM to send multiple signals simultaneously on the same physical waveguide, and optical signals can achieve much higher data rates because they are not limited by capacitance. These facts have led to a lot of research exploring the potential of off-chip photonic interconnects.

For example, in Young et al. [2010], Intel researchers attempt to quantify the pros and cons of using photonics to address the chip I/O problem. They demonstrate a Multi-Chip Module (MCM) solution using polymer waveguides and VCSELs as transmitters, operating at 10Gb/s in 90nm CMOS technology with an energy efficiency of 11pJ/bit. The VCSEL and photodiode arrays are optically coupled to on-package, integrated polymer waveguide arrays using metalized 45° mirrors. The authors estimate that with a transmitter optimized with pre-emphasis circuitry[1] it is possible to achieve an 18Gb/s data rate at 9.6pJ/bit, which can go down to 1pJ/bit at a 16nm technology node. With a monolithic silicon photonics implemention that integrates modulators, waveguides, and detectors on top of metal interconnect layers, they project that a 20Gb/s data rate at an energy efficiency of 0.3pJ/bit is possible.

Columbia researchers are taking a slightly different approach, using microring resonator-based modulators and detectors [Ophir et al., 2013] instead. They propose a system where light is guided by an optical fiber when moving between the transmitter and receiver modules, and by silicon-photonic bus waveguides within the photonic modules themselves. The stated goal of their work is to estimate the achievable energy efficiency in the best case and what needs to happen in order to get to a 1Tb/s data rate with sub-picojoule per-bit efficiency. Essentially they set out to answer the question, what is the trade-off between increasing the number of wavelengths vs. increasing the data rate per wavelength? Their simulation results show that at 12.5Gb/s, the energy

---

[1]This is a technique to reduce inter-symbol interference by emphasizing the high-frequency signal components or attenuating the low-frequency components.

efficiency is 2.5pJ/bit, while at 25Gb/s the energy efficiency falls to 4.3pJ/bit. Their model assumes receivers without amplifiers, a laser efficiency of 10%, and does not take SERDES overhead into account. The drop in efficiency results in large part because higher data rates require tighter channel spacing, which in turn causes higher losses. They also conclude that laser efficiency is a critical factor in the overall wall-plug efficiency of photonic links. This is not a surprise, however—as we have pointed out in this book, the lasers used in telecom applications have incredibly low efficiency (around 1%) because a significant fraction of the power is wasted in stabilizing and cooling the laser cavity, and the insertion losses are quite high as well.

Optical fibers experience extremely low propagation loss (about 0.2dB per kilometer), which makes them very attractive for use in off-chip interconnects. However, the challenge is finding a way to interface tightly coupled waveguide arays (with a 20μm pitch, for example) to standard fiber-optic components (which commonly have a 250μm pitch), especially given the numerical aperture mismatches between silicon waveguides (which have a high index of refraction) and the fibers (which have a lower index). In Lee et al. [2010], IBM researchers propose a multichannel tapered coupler, which can accomplish this goal with coupling losses below 1dB, and cross talk of -35dB between channels at an injection bandwidth of 160Gb/s/channel (4 wavelengths each at 40Gb/s). This could be an important technology down the road and is something that computer architects should pay close attention to, since it might provide a way to integrate many small chips together to form a high-performance system which avoids the thermal constraint problem while simultaneously overcoming the conventional pin limitations of electrical I/O (without compromising latency).

Sun (now Oracle) researchers exploited the unique features of off-chip photonic interconnects when they proposed a *macrochip* [Krishnamoorthy et al., 2009], a logically contiguous piece of photonically interconnected silicon integrating multicore processors, a system-wide interconnect, and dense memory at the wafer level to provide significant improvement in performance and energy efficiency. The logical architecture of a macrochip is based on an 8x8 array of sites where each site contains a four-core processor and 8GB of DRAM, implemented on an SOI platform which contains the CPUs, memory, silicon photonics, and fiber interfaces. The authors propose a fully connected point-to-point interconnect network to support a shared-memory programming model [Koka et al., 2010] initially, and then explore the design space of switched photonic multichip interconnects in subsequent work [Koka et al., 2012]. A bridge chip, which contains the processor and system interface and communicates with the DRAM chip using electrical proximity communication and to waveguides using optical proximity communication, is mounted face down over the DRAM chip. Using this macrochip approach potentially removes the die size limits that constrain the computational efficiency of silicon and could be of interest to future computer architects.

6.2.2 MEMORY SYSTEM

Memory has been a bottleneck in stored-program machines from the dawn of computing, and with the advent of large-scale multicore machines, the so-called memory wall has become even more of a challenge. To keep multiple cores supplied with data, the demands on memory capacity and memory bandwidth have increased dramatically. DDR3 memory channels today provide a peak bandwidth of 25GB/s and support 3 Dual In-line Memory Modules (DIMMs) per channel (or 24GB per channel, assuming 8GB DIMMs). As technology scaling continues, more and more processing cores are being put on the same die, which means these dies will need to be able to support hundreds (or thousands) of GBs of DRAM and a bandwidth of hundreds (or thousands) of GB/s. Traditional DDR3 memory channels require 240 pins, so as noted above, increasing die bandwidth and capacity by adding more channels is not feasible. Furthermore, as the amount of memory increases, the power consumed by that memory becomes a significant component of total system power.

Over the past five years, the use of photonic interconnects to alleviate both the bandwidth and the capacity problem in memory systems has been studied. In Hadke et al. [2008b], Hadke et al. [2008a], and Mejia et al. [2010] the authors propose replacing the electrical store and forward network in a Fully Buffered DIMM (FBDIMM) system with a WDM-based optical bus, in order to improve the bandwidth (and more importantly, the memory capacity) without compromising memory latency.

High-speed multidrop buses are notoriously difficult to build electrically, because of signal integrity issues due to the impedance discontinuities. HP researchers [Tan et al., 2008] have demonstrated a high-speed multidrop bus that uses hollow metal waveguides and pellicle beam splitters (the optical equivalent of electrical stubs in a transmission line) that can connect up to 8 modules at 10Gb/s, with an aggregate bandwidth of 25GB/s (using 10-bit data paths powered by a 1mW laser). This work is intriguing because hollow metal waveguides have very low propagation loss (0.05dB/cm), are easy to fabricate, have low latency (33ps/cm), and their low numerical aperture allows insertions of taps with little loss.

HP [Vantrease et al., 2008] and MIT researchers [Batten et al., 2008] have also shown how an on-chip photonic network can be seamlessly extended to include off-chip memory in order to provide significantly higher bandwidth and memory capacity without having to worry about pin limitations and restrictions on the number of DIMMs per memory channel. However, these works do not explicitly address the potential of exploiting the unique properties of photonics to reduce the power consumption of the DRAM memory system, because that involves re-architecting the DRAM chip itself.

Research presented in Udipi et al. [2011] and Beamer et al. [2010] tries to address this opportunity. In Udipi et al. [2011], the researchers propose a new packet-based memory interface that uses a WDM-based photonic interconnect and relinquishes the tight control that memory controllers hold in current (electrically interconnected) systems, allowing memory modules to be more autonomous. The key enabler is a 3D stacked interface die that supports this new memory

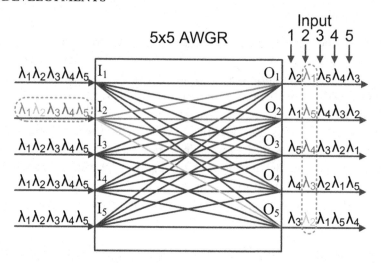

Figure 6.2: All-to-all interconnectivity in 5×5 AWGR.

controller without requiring any modifications to the memory dies themselves. In the Photonically Interconnected DRAM (PIDRAM) described in Beamer et al. [2010], the authors propose a new architecture for a DIMM that takes advantage of monolithically integrated silicon photonics. Specifically, by redesigning DRAM banks to provide greater bandwidth from an individual array core, they can supply the bandwidth demands using much smaller pages and thus reduce bank activation energy. Furthermore, the authors propose split-photonic buses (where optical power is split between multiple direct connections to each bank) and guided-photonic buses (where optical power is actively guided to a single bank). They do this to avoid the use of a shared bus to connect the processor and memory—an optical shared bus does not scale well because broadcasting optical signals requires a significant increase in power.

6.2.3 LARGE-SCALE ROUTERS

In large-scale computer networks such as those employed by supercomputers and datacenters, the use of high-radix switches can reduce latency by reducing the hop count. This in turn could reduce power consumption, because the number of times packets have to be stored and retrieved from power-hungry buffers is also reduced. But as noted earlier in this section, the maximum amount of chip I/O is more or less fixed for the forseeable future, which limits the number of ports of an electrical router. In Binkert et al. [2011], the authors argue that photonics could be used to build high-radix routers—more specifically, they propose an optical switch architecture that exploits high-speed optical interconnects to build a flat crossbar with multiple-writer, single-reader links. The authors suggest that using photonic I/O allows the creation of 100,000 port networks that require only one-third the power of an all-electronic network.

The Arrayed Waveguide Grating Router (AWGR) [Glance et al., 1994; Takada et al., 2001] is a well-understood optical component that can be used to implement a highly scalable, non-blocking optical switch matrix. It has been used in many packet switch architectures for telecom applications [Yoo, 2006], and has proven to be very useful for creating low-latency switches in datacenter applications [Ye et al., 2010]. The AWGR allows a signal from one input to reach exactly one output on a particular wavelength. Figure 6.2 shows a wavelength routing map of a 5×5 AWGR with 5 wavelengths per port. At AWGR input 2, it is possible to use $\lambda_1, \lambda_2, \lambda_3, \lambda_4,$ and λ_5 to reach outputs 1, 5, 4, 3, and 2 respectively; one can also use $\lambda_4, \lambda_3, \lambda_2, \lambda_1,$ and λ_5 to reach AWGR output 4 from inputs 1, 2, 3, 4, and 5, respectively. Note that wavelength routing in AWGR exhibits a symmetric characteristic, for example λ_2 can be used to route from input 2 to output 5 and from input 5 to output 2. The AWGR is interesting in that all-to-all communication can be realized if every node is equipped with multiple receivers and multiple transmitters working on different wavelengths, so that signals on different wavelengths can be transmitted and received concurrently. Furthermore, the AWGR fabric is passive and low-loss at all data rates, which is a significant advantage in itself—power consumption is a major obstacle to scalability in computing, especially at high speeds. In Proietti et al. [2012a,b], the authors demonstrate efficient contention-resolution techniques that take advantage of an Reflective Semiconductor Optical Amplifier (RSOA) to implement mutual exclusion in the optical domain. Recent work by the same research group has shown that it is possible to realize chip-scale optical crossconnects using AWGR which, together with low-latency contention resolution at the physical layer, offers new opportunities for on-chip photonic networks in the future.

CHAPTER 7

Summary & Conclusion

As the number of cores on a chip continues to climb, architects will need to address both bandwidth and power consumption issues related to the on-chip interconnection network. The problem is rooted in the fact that there is a fixed total power budget for a die, dictated by the amount of heat that can be dissipated without the use of special (and expensive) cooling and packaging techniques. With each generation of electrical interconnects consuming more power to provide the necessary increase in bandwidth, there is less energy available for *processing*. Thus, it is important that architects begin to look elsewhere for solutions.

Photonics, with a fundamentally different mechanism of signal propagation, has the potential to overcome the drawbacks of electrical signaling. In addition, it offers opportunities to build scalable systems that are energy-efficient and perhaps easier to program. The purpose of this short book has been to introduce computer architects to the possibilities and design trade-offs involved when using photonic interconnects in on-chip applications. In this final chapter we will provide a recap of what we think are some the most important take-away messages presented in this book. (The interested reader may also find the following on-chip nanophotonic survey papers useful: Beausoleil et al. [2008], Beausoleil [2011], and Batten et al. [2012].)

7.1 OBSERVATIONS AND THINGS TO REMEMBER

Observation: Electrons vs. Photons

Loosely speaking, in electrical signaling data is communicated along a wire by charging or discharging a remote node. The resistance of both the driver and the wire, as well as the capacitance of the target node, influence the delay characteristics of electrical signal propagation. As wire geometries shrink and the relative distance between the sender and the receiver grows, the contribution of the wire resistance to both the delay and the power consumption becomes increasingly problematic. Photonics, on the other hand, does not have this problem because data is communicated by modulating photons that move along a waveguide. While there are losses in the waveguide, they are not as deleterious as the resistance of the wire. In fact, the maturation of fabrication processes that enable the creation of waveguides and microring resonators is a key reason why photonic interconnects are now conceivable for use in on-chip applications.

However, there are a couple of key differences between electrons and photons that the architect needs to keep in mind, because they directly impact network design. One is the fact that photons cannot be stored easily (if at all). Any network which requires buffering is going to have to convert photons to electrons and back, which is a slow and expensive process. Another important difference is that broadcasting a photonic signal is not as straightforward as it is using electronics, and requires a large amount of power.

Remember: Electronic/Photonic Interconnect Codesign

Despite its various drawbacks, an electrical interconnect does have certain advantages—first, it is very efficient for short-distance communication in terms of delay and energy, and second, signals can be buffered very easily with the buffers themselves becoming smaller and more efficient as technology scales. Photons, on the other hand, are good for longer-distance communication, but buffering (or temporarily storing) photons is impractical, if not impossible. As we described in Chapter 3 and Chapter 4, the partition-length is a guide to choosing when electronics are preferable to photonics. What this means is that the design of an on-chip network has to be posed as a codesign problem that balances the unique advantages and disadvantages of both electronics and photonics in order to meet the desired bandwidth and power constraints.

Remember: Bandwidth Limits with WDM and Microring Resonators

As noted in Chapter 3, there is a fundamental limit to how small microring resonators can be made, so as electronics continue to shrink but the ring dimensions do not, the power consumed by the electrical transport network needed to get the signals to the microrings becomes a significant factor. This makes increasing the bandwidth of a link that uses DWDM challenging—adding wavelengths requires the use of more rings, which means an increase in the distance the signals in the electrical transport network have to travel, adding to the latency and power penalty of the link. In addition, there is a limit to the amount of DWDM that can be used, due to signal cross-talk issues. Furthermore, when using off-chip lasers the photonic power entering the chip is fixed—it is a pre-paid cost, independent of whether or not data is actually communicated using those incoming photons. Thus, the cost of unused bandwidth becomes an important issue to consider as well, which means that, in addition to the partition-length, link utilization is also an important factor in electronic/photonic interconnect codesign. Simply replacing individual electronic links with photonic links is unlikely to produce the desired results, and a system-level approach to network design is required.

Observation: Laser Efficiency Must Improve

Current on-chip network designs assume the necessary photons are created externally, and standard off-the-shelf lasers are used so that researchers can focus on the things that are unique to on-chip photonics. However, the efficiency of some standard off-chip lasers is in the neighborhood of 1%, and even the best lasers fall short of being 30% efficient. In order for on-chip photonic networks to become commercially viable, advances must be made in laser efficiency. Ideally lasers will not only be made more efficient, but they will also be moved on-chip so that they can be turned off when communication is not occuring.

Remember: Thermals and Resilience

As described in Chapter 5, many of the components used in photonic interconnects are extremely sensitive to variations in fabrication and temperature. This is of particular importance for on-chip applications, where not only do we anticipate having hundreds of thousands of these components (in the form of modulators and filters) operating simultaneously, but also in close proximity to heat-generating processors. Clearly, resilience should not be an *afterthought* when using photonic interconnects, but rather should be considered as part of the design itself. This is important to the architect because it impacts things like the choice of modulating zeros or ones—if modulating ones, it is possible to detect and correct all possible single-bit errors. We described some preliminary approaches to fault modeling and protection using simple codes in Chapter 5, but clearly this is an important area that requires serious attention before photonic interconnects can be used widely.

Remember: Importance of Assumptions

As discussed in Chapter 2 and Chapter 3, the power consumption of a photonic link (and thus the power that must be supplied by the laser) is determined by starting with the minimum energy required by the receiver to detect a photon and working backward toward the laser, taking into account various intermediate losses. If the detector used is not efficient, it will clearly increase the power required for photonic communication. The high efficiency of photonic interconnects in terms of energy per bit reported in many research publications is based on the assumption that this receiver can be made so compact, and hence efficient, that an amplifier (such as a TIA) is not needed. Developing such receivers and making sure that they function properly in the large numbers required for an on-chip application is a key underlying (and often understated) assumption made in on-chip photonics interconnects literature.

In fact, the attenuation values used by different researchers for the components of optical links vary quite dramatically (as noted in Chapter 3). Designers should be very careful when choosing loss values from the literature, and whenever possible architects should work with device experts to get a *consistent* set of attenuation values—they should all be optimistic or conservative, and all assume the same technology.

Observation: Parallel Programming and Photonic Interconnects

The use of photonics also offers the potential to remove or at least minimize the dichotomy between on-chip and off-chip communication, a problem that has plagued electrical signaling from the beginning. This, coupled with the ability to communicate over longer distance with a much lower penalty, offers opportunities for new types of computing devices—architectures that are based on flatter communication hierarchies and have close to uniform communication costs, even when systems are scaled to hundreds of thousands of processing nodes. Can these new architectures improve programmer productivity? What is the design space of such architectures? We think these are interesting questions that require further research, especially in the context of exascale computing systems of the future.

Bibliography

Jung H. Ahn, Marco Fiorentino, Raymond G. Beausoleil, Nathan Binkert, Al Davis, David Fattal, Norman P. Jouppi, Moray McLaren, Charles M. Santori, Robert S. Schreiber, Sean M. Spillane, Dana Vantrease, and Qianfan Xu. Devices and architectures for photonic chip-scale integration. *Applied Physics A: Materials Science & Processing*, 95:989–997, June 2009. DOI: 10.1007/s00339-009-5109-2. 16, 22, 23, 41, 53

Baba Arimilli, Ravi Arimilli, Vicente Chung, Scott Clark, Wolfgang Denzel, Ben Drerup, Torsten Hoefler, Jody Joyner, Jerry Lewis, Jian Li, Nan Ni, and Ram Rajamony. The PERCS high-performance interconnect. *High-Performance Interconnects, Symposium on*, 0:75–82, 2010. DOI: 10.1109/HOTI.2010.16. 2

Christopher Batten, Ajay Joshi, Jason Orcutt, Anatoly Khilo, Benjamin Moss, Charles Holzwarth, Milos Popovic, Hanqing Li, Henry Smith, Judy Hoyt, Franz Kartner, Rajeev Ram, Vladimir Stojanović, and Krste Asanović. Building manycore processor-to-DRAM networks with monolithic silicon photonics. In *HOTI '08: Proceedings of the 2008 16th IEEE Symposium on High Performance Interconnects*, pages 21–30, Washington, DC, USA, 2008. IEEE Computer Society. ISBN 978-0-7695-3380-3. DOI: 10.1109/HOTI.2008.11. 69

Christopher Batten, Ajay Joshi, Vladimir Stojanović, and Krste Asanović. Designing chip-level nanophotonic interconnection networks. *Emerging and Selected Topics in Circuits and Systems, IEEE Journal on*, 2(2):137–153, 2012. ISSN 2156-3357. DOI: 10.1109/JET-CAS.2012.2193932. 73

Scott Beamer, Chen Sun, Yong-Jin Kwon, Ajay Joshi, Christopher Batten, Vladimir Stojanović, and Krste Asanović. Re-architecting DRAM memory systems with monolithically integrated silicon photonics. In *Proceedings of the 37th annual International Symposium on Computer Architecture*, ISCA '10, pages 129–140, New York, NY, USA, 2010. ACM. ISBN 978-1-4503-0053-7. DOI: 10.1145/1816038.1815978. 69, 70

Raymond G. Beausoleil. Large-scale integrated photonics for high-performance interconnects. *Journal on Emerging Technologies in Computing Systems*, 7(2):6:1–6:54, July 2011. ISSN 1550-4832. URL http://doi.acm.org/10.1145/1970406.1970408. DOI: 10.1109/PHO.2011.6110559. 64, 73

Raymond G. Beausoleil, Phillip J. Kuekes, Gregory S. Snider, Shih-Yuan Wang, and Richard Stanley Williams. Nanoelectronic and nanophotonic interconnect. *Pro-*

ceedings of the IEEE, 96(2):230–247, February 2008.   ISSN 0018-9219. DOI: 10.1109/JPROC.2007.911057. 32, 73

Alan F. Benner, Michael Ignatowski, Jeffrey A. Kash, Daniel M. Kuchta, and Mark B. Ritter. Exploitation of optical interconnects in future server architectures. *IBM Journal of Research and Development*, 49(4/5):755–775, 2005. ISSN 0018-8646. DOI: 10.1147/rd.494.0755. 2

Jay M. Berger. A note on error detection codes for asymmetric channels. *Information and Control*, 4(1):68 – 73, 1961. ISSN 0019-9958. URL http://www.sciencedirect.com/science/article/pii/S0019995861800375. DOI: 10.1016/S0019-9958(61)80037-5. 59

Ryan Douglas Bespalko. Transimpedance amplifier design using 0.18μm CMOS technology. Master's thesis, Queen's University, 2007. 17

Apama Bhatnagar, Christof Debaes, Ray Chen, Noah C. Hellman, Gordon A. Keeler, Diwakar Agarwal, Hugo Thienpont, and David A. B. Miller. Receiverless clocking of a CMOS digital circuit using short optical pulses. In *Lasers and Electro-Optics Society, 2002. LEOS 2002. The 15th Annual Meeting of the IEEE*, volume 1, pages 127–128 vol.1, 2002. DOI: 10.1109/LEOS.2002.1133951. 16

Nathan Binkert, Al Davis, Norman P. Jouppi, Moray McLaren, Naveen Muralimanohar, Robert Schreiber, and Jung Ho Ahn. The role of optics in future high radix switch design. In *Proceedings of the 38th Annual International Symposium on Computer Architecture*, ISCA '11, pages 437–448, New York, NY, USA, 2011. ACM. ISBN 978-1-4503-0472-6. DOI: 10.1145/2024723.2000116. 70

Ozdal Boyraz and Bahram Jalali. Demonstration of a silicon Raman laser. *Optics Express*, 12(21): 5269–5273, Oct 2004. URL http://www.opticsexpress.org/abstract.cfm?URI=oe-12-21-5269. DOI: 10.1364/OPEX.12.005269. 64

Wenshan Cai, Justin S. White, and Mark L. Brongersma. Compact, high-speed and power-efficient electrooptic plasmonic modulators. *Nano Letters*, 9(12):4403–4411, 2009. URL http://pubs.acs.org/doi/abs/10.1021/nl902701b. PMID: 19827771. DOI: 10.1021/nl902701b. 65

Rodolfo E. Camacho-Aguilera, Yan Cai, Neil Patel, Jonathan T. Bessette, Marco Romagnoli, Lionel C. Kimerling, and Jurgen Michel. An electrically pumped germanium laser. *Optics Express*, 20(10):11316–11320, May 2012. URL http://www.opticsexpress.org/abstract.cfm?URI=oe-20-10-11316. DOI: 10.1364/OE.20.011316. 64

Giuseppe Campobello, Giuseppe Patane, and Marco Russo.   Parallel CRC realization. *IEEE Transactions on Computers*, 52:1312–1319, 2003.   ISSN 0018-9340. DOI: 10.1109/TC.2003.1234528. 59

Linyou Cao, Joon-Shik Park, Pengyu Fan, Bruce Clemens, and Mark L. Brongersma. Resonant germanium nanoantenna photodetectors. *Nano Letters*, 10(4):1229–1233, 2010. URL http://pubs.acs.org/doi/abs/10.1021/nl9037278. PMID: 20230043. DOI: 10.1021/nl9037278. 16, 66

Mark J. Cianchetti, Joseph C. Kerekes, and David H. Albonesi. Phastlane: a rapid transit optical routing network. In *Proceedings of the 36th Annual International Symposium on Computer Architecture*, ISCA '09, pages 441–450, New York, NY, USA, 2009. ACM. ISBN 978-1-60558-526-0. URL http://doi.acm.org/10.1145/1555754.1555809. DOI: 10.1145/1555815.1555809. 36, 41, 42

Tibor Cinkler. Traffic and lambda; grooming. *Network, IEEE*, 17(2):16–21, 2003. ISSN 0890-8044. DOI: 10.1109/MNET.2003.1188282. 38

Paul W. Coteus, John U. Knickerbocker, Chung H. Lam, and Yuri A. Vlasov. Technologies for exascale systems. *IBM Journal of Research and Development*, 55(5):14:1–14:12, 2011. ISSN 0018-8646. DOI: 10.1147/JRD.2011.2163967. 2

William Dally and Brian Towles. *Principles and Practices of Interconnection Networks*. Morgan Kaufmann, San Francisco, 2004. ISBN 978-0-12-200751-4. 36

Jennifer A. Dionne, Kenneth Diest, Luke A. Sweatlock, and Harry A. Atwater. PlasMOStor: A Metal-Oxide-Si Field Effect Plasmonic Modulator. *Nano Letters*, 9(2):897–902, 2009. URL http://pubs.acs.org/doi/abs/10.1021/nl803868k. PMID: 19170558. DOI: 10.1021/nl803868k. 65

Jennifer A. Dionne, Luke A. Sweatlock, Matthew T. Sheldon, A. Paul Alivisatos, and Harry A. Atwater. Silicon-based plasmonics for on-chip photonics. *Selected Topics in Quantum Electronics, IEEE Journal of*, 16(1):295 –306, Jan.-Feb. 2010. ISSN 1077-260X. DOI: 10.1109/JSTQE.2009.2034983. 65

Nathan Farrington, George Porter, Sivasankar Radhakrishnan, Hamid Hajabdolali Bazzaz, Vikram Subramanya, Yeshaiahu Fainman, George Papen, and Amin Vahdat. Helios: a hybrid electrical/optical switch architecture for modular data centers. In *Proceedings of the ACM SIGCOMM 2010 conference*, SIGCOMM '10, pages 339–350, New York, NY, USA, 2010. ACM. ISBN 978-1-4503-0201-2. URL http://doi.acm.org/10.1145/1851182.1851223. DOI: 10.1145/1851275.1851223. 2, 38

Ning-Ning Feng, Shirong Liao, Dazeng Feng, Po Dong, Dawei Zheng, Hong Liang, Roshanak Shafiiha, Guoliang Li, John E. Cunningham, Ashok V. Krishnamoorthy, and Mehdi Asghari. High speed silicon carrier-depletion Mach-Zehnder modulator with 1.4V-cm VπL. In *Integrated Photonics Research, Silicon and Nanophotonics*, page IMB3. Optical Society of America, 2010. URL http://www.opticsinfobase.org/abstract.cfm?URI=IPRSN-2010-IMB3. 12

Mark A. Foster, Jacob S. Levy, Onur Kuzucu, Kasturi Saha, Michal Lipson, and Alexander L. Gaeta. Silicon-based monolithic optical frequency comb source. *Optics Express*, 19(15):14233–14239, July 2011. URL http://www.opticsexpress.org/abstract.cfm?URI=oe-19-15-14233. DOI: 10.1364/OE.19.014233. 65

Bradley J. Frey, Douglas B. Leviton, and Timothy J. Madison. Temperature-dependent refractive index of silicon and germanium. In *Society of Photo-Optical Instrumentation Engineers (SPIE) Conference Series*, volume 6273 of *Society of Photo-Optical Instrumentation Engineers (SPIE) Conference Series*, July 2006. DOI: 10.1117/12.672850. 53

Michael Georgas, Jonathan Leu, Benjamin Moss, Chen Sun, and Vladimir Stojanović. Addressing link-level design tradeoffs for integrated photonic interconnects. In *Custom Integrated Circuits Conference (CICC), 2011 IEEE*, pages 1–8, 2011. DOI: 10.1109/CICC.2011.6055363. 16, 17, 34

Bernard Glance, Ivan P. Kaminow, and Robert W. Wilson. Applications of the integrated waveguide grating router. *Lightwave Technology, Journal of*, 12(6):957–962, 1994. ISSN 0733-8724. DOI: 10.1109/50.296184. 71

Alexey Gubenko, Igor Krestnikov, Daniil Livshtis, Sergey Mikhrin, Alexey Kovsh, Lawrence West, Carsten Bornholdt, Norbert Grote, and Alexey Zhukov. Error-free 10 Gbit/s transmission using individual Fabry-Perot modes of low-noise quantum-dot laser. *Electronics Letters*, 43(25):1430–1431, 2007. ISSN 0013-5194. DOI: 10.1049/el:2007295. 64

Biswajeet Guha, Bernardo B. C. Kyotoku, and Michal Lipson. CMOS-compatible athermal silicon microring resonators. *Optics Express*, 18(4):3487–3493, February 2010. URL http://www.opticsexpress.org/abstract.cfm?URI=oe-18-4-3487. DOI: 10.1364/OE.18.003487. 52

Amit Hadke, Tony Benavides, Rajeevan Amirtharajah, Matthew K. Farrens, and Venkatesh Akella. Design and evaluation of an optical CPU-DRAM interconnect. In *Computer Design, 2008. ICCD 2008. IEEE International Conference on*, pages 492–497, October 2008a. DOI: 10.1109/ICCD.2008.4751906. 69

Amit Hadke, Tony Benavides, S. J. Ben Yoo, Rajeevan Amirtharajah, and Venkatesh Akella. OCDIMM: Scaling the DRAM memory wall using WDM based optical interconnects. In *High Performance Interconnects, 2008. HOTI '08. 16th IEEE Symposium on*, pages 57 –63, August 2008b. DOI: 10.1109/HOTI.2008.25. 69

B. Roe Hemenway. Roadmap for Board Level Optical Interconnects – They're Coming Sooner Than You Think!, July 2013. URL http://www.semiconwest.org/sites/semiconwest.org/files/docs/SW2013_Roe%20%20Hemenway_Photonic%20Controls.pdf. Invited talk, SEMICON West 2013. 2

Gilbert Hendry, Shoaib Kamil, Aleksandr Biberman, Johnnie Chan, Benjamin G. Lee, Marghoob Mohiyuddin, Ankit Jain, Keren Bergman, Luca P. Carloni, John Kubiatowicz, Leonid Oliker, and John Shalf. Analysis of photonic networks for a chip multiprocessor using scientific applications. In *Networks-on-Chip, 2009. NoCS 2009. 3rd ACM/IEEE International Symposium on*, pages 104–113, Los Alamitos, CA, USA, 2009. IEEE Computer Society. ISBN 978-1-4244-4142-6. DOI: 10.1109/NOCS.2009.5071458. 50

Intel. Enhanced Intel® SpeedStep® Technology for the Intel® Pentium® M Processor. Technical report, Intel Corporation, March 2004. URL `ftp://download.intel.com/design/network/papers/30117401.pdf`. Available online (12 pages). 23

Intel. Measuring processor power: TDP vs. ACP. Technical report, Intel Corporation, April 2011. URL `http://www.intel.com/content/dam/doc/white-paper/resources-xeon-measuring-processor-power-paper.pdf`. Available online (8 pages). 23

Bahram Jalali and Sasan Fathpour. Silicon photonics. *Lightwave Technology, Journal of*, 24(12): 4600–4615, 2006. ISSN 0733-8724. 64

Joe Jeddeloh and Brent Keeth. Hybrid memory cube new DRAM architecture increases density and performance. In *VLSI Technology (VLSIT), 2012 Symposium on*, pages 87–88, 2012. DOI: 10.1109/VLSIT.2012.6242474. 4

Ajay Joshi, Christopher Batten, Yong-Jin Kwon, Scott Beamer, Imran Shamim, Krste Asanović, and Vladimir Stojanović. Silicon-photonic clos networks for global on-chip communication. In *NOCS '09: Proceedings of the 2009 3rd ACM/IEEE International Symposium on Networks-on-Chip*, pages 124–133, Washington, DC, USA, 2009a. IEEE Computer Society. ISBN 978-1-4244-4142-6. DOI: 10.1109/NOCS.2009.5071460. 15, 23, 26, 48, 53

Ajay Joshi, Byungsub Kim, and Vladimir Stojanović. Designing energy-efficient low-diameter on-chip networks with equalized interconnects. In *High Performance Interconnects, 2009. HOTI 2009. 17th IEEE Symposium on*, pages 3–12, Los Alamitos, CA, USA, 2009b. IEEE Computer Society. DOI: 10.1109/HOTI.2009.13. 33

Robert W. Kelsall. Hybrid silicon lasers: Rubber stamp for silicon photonics. *Nature Photonics*, 6(9):577–579, 2012. DOI: 10.1038/nphoton.2012.211. 64

Byungsub Kim and Vladimir Stojanović. Future state-of-the-art electrical interconnect. In *WINDS 2010 Workshop on the Interaction between Nanophotonic Devices and Systems*, pages 5–6, December 2010. 32

John Kim, Wiliam J. Dally, Steve Scott, and Dennis Abts. Technology-driven, highly-scalable dragonfly topology. In *Proceedings of the 35th Annual International Symposium on Computer*

Architecture, ISCA '08, pages 77–88, Washington, DC, USA, 2008. IEEE Computer Society. ISBN 978-0-7695-3174-8. URL http://dx.doi.org/10.1109/ISCA.2008.19. DOI: 10.1109/ISCA.2008.19. 2

John Kim, William J. Dally, Steve Scott, and Dennis Abts. Cost-efficient dragonfly topology for large-scale systems. *IEEE Micro*, 29(1):33–40, January 2009. ISSN 0272-1732. URL http://dx.doi.org/10.1109/MM.2009.5. DOI: 10.1109/MM.2009.5. 2

Nevin Kirman, Meyrem Kirman, Rajeev K. Dokania, Jose F. Martinez, Alyssa B. Apsel, Matthew A. Watkins, and David H. Albonesi. Leveraging optical technology in future bus-based chip multiprocessors. In *MICRO 39: Proceedings of the 39th Annual IEEE/ACM International Symposium on Microarchitecture*, pages 492–503, Washington, DC, USA, 2006. IEEE Computer Society. ISBN 0-7695-2732-9. DOI: 10.1109/MICRO.2006.28. 17, 41

Brian R. Koch, Alexander W. Fang, Oded Cohen, and John E. Bowers. Mode-locked silicon evanescent lasers. *Optics Express*, 15(18):11225–11233, September 2007. URL http://www.opticsexpress.org/abstract.cfm?URI=oe-15-18-11225. DOI: 10.1364/OE.15.011225. 8

Thomas L. Koch. Opportunities and challenges in silicon photonics. In *Lasers and Electro-Optics Society, 2006. LEOS 2006. 19th Annual Meeting of the IEEE*, pages 677 –678, October 2006. DOI: 10.1109/LEOS.2006.278894. 26

Pranay Koka, Michael O. McCracken, Herb Schwetman, Xuezhe Zheng, Ron Ho, and Ashok V. Krishnamoorthy. Silicon-photonic network architectures for scalable, power-efficient multi-chip systems. In *Proceedings of the 37th Annual International Symposium on Computer Architecture*, ISCA '10, pages 117–128, New York, NY, USA, 2010. ACM. ISBN 978-1-4503-0053-7. URL http://doi.acm.org/10.1145/1815961.1815977. DOI: 10.1145/1816038.1815977. 68

Pranay Koka, Michael O. McCracken, Herb Schwetman, Chia-Hsin Owen Chen, Xuezhe Zheng, Ron Ho, Kannan Raj, and Ashok V. Krishnamoorthy. A micro-architectural analysis of switched photonic multi-chip interconnects. In *Proceedings of the 39th International Symposium on Computer Architecture*, ISCA '12, pages 153–164, Piscataway, NJ, USA, 2012. IEEE Press. ISBN 978-1-4503-1642-2. DOI: 10.1109/ISCA.2012.6237014. 15, 68

Yasuo Kokubun, Naoki Kobayashi, and Tomoyuki Sato. UV trimming of polarization-independent microring resonator by internal stress and temperature control. *Optics Express*, 18(2):906–916, 2010. DOI: 10.1364/OE.18.000906. 51

Ashok V. Krishnamoorthy, Ron Ho, Xuezhe Zheng, Herb Schwetman, Jon Lexau, Pranay Koka, Guoliang Li, Ivan Shubin, and John E. Cunningham. Computer systems based on silicon photonic interconnects. *Proceedings of the IEEE*, 97(7):1337–1361, 2009. ISSN 0018-9219. DOI: 10.1109/JPROC.2009.2020712. 68

Benjamin G. Lee, Fuad E. Doany, Solomon Assefa, William M. J. Green, Min Yang, Clint L. Schow, Christopher V. Jahnes, Sheng Zhang, Jonathan Singer, Victor I. Kopp, Jeffrey A. Kash, and Yurii A. Vlasov. 20-μm-pitch eight-channel monolithic fiber array coupling 160 Gb/s/channel to silicon nanophotonic chip. In *Optical Fiber Communication (OFC), collocated National Fiber Optic Engineers Conference, 2010 Conference on (OFC/NFOEC)*, pages 1–3, 2010. DOI: 10.1364/OFC.2010.PDPA4. 68

Jacob S. Levy, Alexander Gondarenko, Mark A. Foster, Amy C. Turner-Foster, Alexander L. Gaeta, and Michal Lipson. CMOS-compatible multiple-wavelength oscillator for on-chip optical interconnects. *Nature Photonics*, 4(1):37–40, 2009. DOI: 10.1038/nphoton.2009.259. 63

Di Liang and John E Bowers. Recent progress in lasers on silicon. *Nature Photonics*, 4(8):511–517, 2010. DOI: 10.1038/nphoton.2010.167. 64

Michal Lipson. Guiding, modulating, and emitting light on silicon-challenges and opportunities. *Lightwave Technology, Journal of*, 23(12):4222–4238, December 2005. ISSN 0733-8724. DOI: 10.1109/JLT.2005.858225. 7, 15, 51, 52

Hong Liu, Cedric F. Lam, and Chris Johnson. Scaling optical interconnects in datacenter networks opportunities and challenges for WDM. In *High Performance Interconnects (HOTI), 2010 IEEE 18th Annual Symposium on*, pages 113–116, 2010a. DOI: 10.1109/HOTI.2010.15. 8

Jifeng Liu, Xiaochen Sun, Rodolfo Camacho-Aguilera, Lionel C. Kimerling, and Jurgen Michel. Ge-on-Si laser operating at room temperature. *Optics Letters*, 35(5):679–681, March 2010b. URL http://ol.osa.org/abstract.cfm?URI=ol-35-5-679. DOI: 10.1364/OL.35.000679. 64

Weidong Lu and Stephan Wong. A fast CRC update implementation. In *Proceedings of the 14th Annual Workshop on Circuits, Systems and Signal Processing*, pages 113–120, November 2003. 59

Guillaume Maire, Laurent Vivien, Guillaume Sattler, Andrzej Kazmierczak, Benito Sanchez, Kristinn B. Gylfason, Amadeu Griol, Delphine Marris-Morini, Eric Cassan, Domenico Giannone, Hans Sohlström, and Daniel Hill. High efficiency silicon nitride surface grating couplers. *Optics Express*, 16(1):328–333, 2008. URL http://www.opticsexpress.org/abstract.cfm?URI=oe-16-1-328. DOI: 10.1364/OE.16.000328. 15

Ivan S. Maksymov, Isabelle Staude, Andrey E. Miroshnichenko, and Yuri S. Kivshar. Optical Yagi-Uda nanoantennas. *Nanophotonics*, 1:65–81, July 2012. DOI: 10.1515/nanoph-2012-0005. 66

Paul V. Mejia, Rajeevan Amirtharajah, Matthew K. Farrens, and Venkatesh Akella. Performance evaluation of a multicore system with optically connected memory modules. In *Networks-on-Chip (NOCS), 2010 Fourth ACM/IEEE International Symposium on*, pages 215 –222, May 2010. DOI: 10.1109/NOCS.2010.31. 69

David A. B. Miller. Optics for low-energy communication inside digital processors: quantum detectors, sources, and modulators as efficient impedance converters. *Optics Letters*, 14(2):146–148, January 1989. URL http://ol.osa.org/abstract.cfm?URI=ol-14-2-146. DOI: 10.1364/OL.14.000146. 17

David A. B. Miller. Rationale and challenges for optical interconnects to electronic chips. *Proceedings of the IEEE*, 88(6):728–749, June 2000. ISSN 0018-9219. DOI: 10.1109/5.867687. 4

David A. B. Miller. Device requirements for optical interconnects to silicon chips. *Proceedings of the IEEE*, 97(7):1166 –1185, July 2009. ISSN 0018-9219. DOI: 10.1109/JPROC.2009.2014298. 4, 8, 16, 17

Makoto Motoyoshi. Through-Silicon Via (TSV). *Proceedings of the IEEE*, 97(1):43–48, 2009. ISSN 0018-9219. 4

Azad Naeemi, Jianping Xu, Anthony V. Mulé, Thomas K. Gaylord, and James D. Meindl. Optical and electrical interconnect partition length based on chip-to-chip bandwidth maximization. *Photonics Technology Letters, IEEE*, 16(4):1221 –1223, April 2004. ISSN 1041-1135. DOI: 10.1109/LPT.2004.824623. 27, 32

Christopher Nitta, Matthew Farrens, and Venkatesh Akella. DCOF - An arbitration free directly connected optical fabric. *Emerging and Selected Topics in Circuits and Systems, IEEE Journal on*, 2(2):169 –182, June 2012a. ISSN 2156-3357. DOI: 10.1109/JETCAS.2012.2193842. 22, 48

Christopher Nitta, Matthew Farrens, and Venkatesh Akella. Evaluating the energy efficiency of microring resonator-based on-chip photonic interconnects. Technical Report CSE-2012-35, Department of Computer Science, University of California, Davis, 2012b. 31, 32

Christopher J. Nitta. *Design and Analysis of Large Scale Nanophotonic On-Chip Networks*. PhD thesis, University of California, Davis, 2011. 16, 46, 48, 59, 61

Christopher J. Nitta, Matthew K. Farrens, and Venkatesh Akella. Addressing system-level trimming issues in on-chip nanophotonic networks. In *High Performance Computer Architecture (HPCA), 2011 IEEE 17th International Symposium on*, pages 122–131, February 2011a. DOI: 10.1109/HPCA.2011.5749722. 24, 53

Christopher J. Nitta, Matthew K. Farrens, and Venkatesh Akella. Resilient microring resonator based photonic networks. In *Proceedings of the 44th Annual IEEE/ACM International Symposium on Microarchitecture*, MICRO-44 '11, pages 95–104, New York, NY, USA, 2011b. ACM. ISBN 978-1-4503-1053-6. DOI: 10.1145/2155620.2155632. 54

Christopher J. Nitta, Matthew K. Farrens, and Venkatesh Akella. DCAF - A directly connected arbitration-free photonic crossbar for energy-efficient high performance computing. In *26th IEEE International Parallel and Distributed Processing Symposium*, pages 1144–1155, May 2012c. DOI: 10.1109/IPDPS.2012.105. 16, 21, 36, 46, 48

Mike J. O'Mahony, Dimitra Simeonidou, David K. Hunter, and Anna Tzanakaki. The application of optical packet switching in future communication networks. *Communications Magazine, IEEE*, 39(3):128–135, 2001. ISSN 0163-6804. DOI: 10.1109/35.910600. 38

Noam Ophir, Christopher Mineo, David Mountain, and Keren Bergman. Silicon photonic microring links for high-bandwidth-density, low-power chip I/O. *Micro, IEEE*, 33(1):54–67, 2013. ISSN 0272-1732. DOI: 10.1109/MM.2013.1. 34, 67

Jason S. Orcutt and Rajeev J. Ram. Photonic device layout within the foundry CMOS design environment. *Photonics Technology Letters, IEEE*, 22(8):544–546, 2010. ISSN 1041-1135. DOI: 10.1109/LPT.2010.2041445. 63

Jason S. Orcutt, Anatol Khilo, Charles W. Holzwarth, Milos A. Popović, Hanqing Li, Jie Sun, Thomas Bonifield, Randy Hollingsworth, Franz X. Kärtner, Henry I. Smith, Vladimir Stojanović, and Rajeev J. Ram. Nanophotonic integration in state-of-the-art CMOS foundries. *Optics Express*, 19(3):2335–2346, January 2011. URL http://www.opticsexpress.org/abstract.cfm?URI=oe-19-3-2335. DOI: 10.1364/OE.19.002335. 63

Jason S. Orcutt, Benjamin Moss, Chen Sun, Jonathan Leu, Michael Georgas, Jeffrey Shainline, Eugen Zgraggen, Hanqing Li, Jie Sun, Matthew Weaver, Stevan Urošević, Miloš Popović, Rajeev J. Ram, and Vladimir Stojanović. Open foundry platform for high-performance electronic-photonic integration. *Opt. Express*, 20(11):12222–12232, May 2012. URL http://www.opticsexpress.org/abstract.cfm?URI=oe-20-11-12222. DOI: 10.1364/OE.20.012222. 64

Yan Pan, Prabhat Kumar, John Kim, Gokhan Memik, Yu Zhang, and Alok Choudhary. Firefly: illuminating future network-on-chip with nanophotonics. In *Proceedings of the 36th annual international symposium on Computer architecture*, ISCA '09, pages 429–440, New York, NY, USA, 2009. ACM. ISBN 978-1-60558-526-0. URL http://doi.acm.org/10.1145/1555754.1555808. DOI: 10.1145/1555815.1555808. 44, 45, 56

Yan Pan, John Kim, and Gokhan Memik. Flexishare: Channel sharing for an energy-efficient nanophotonic crossbar. In *High Performance Computer Architecture, 2010.*

HPCA 2010. IEEE 16th International Symposium on, pages 1–12, January 2010. DOI: 10.1109/HPCA.2010.5416626. 15, 23, 26, 45, 46, 53

Yan Pan, John Kim, and Gokhan Memik.   FeatherWeight: low-cost optical arbitration with QoS support.   In *Proceedings of the 44th Annual IEEE/ACM International Symposium on Microarchitecture*, MICRO-44 '11, pages 105–116, New York, NY, USA, 2011. ACM.   ISBN 978-1-4503-1053-6.   URL http://doi.acm.org/10.1145/2155620.2155633. DOI: 10.1145/2155620.2155633. 45

Anand M. Pappu and Alyssa B. Apsel.   A low-power, low-delay TIA for on-chip applications.   In *Conference on Lasers and Electro-Optics/Quantum Electronics and Laser Science and Photonic Applications Systems Technologies*, pages 594–596. Optical Society of America, 2005. URL http://www.opticsinfobase.org/abstract.cfm?URI=CLEO-2005-CMGG6. DOI: 10.1109/CLEO.2005.201858. 17

Behrooz Parhami. A multi-level view of dependable computing. *Computers & Electrical Engineering*, 20(4):347 – 368, 1994. ISSN 0045-7906. URL http://www.sciencedirect.com/science/article/pii/0045790694900485. 54

Roberto Proietti, Christopher Nitta, Yawei Yin, Runxiang Yu, S. J. B. Yoo, and Venkatesh Akella. Scalable and distributed contention resolution in AWGR-based data center switches using RSOA-based optical mutual exclusion. *Selected Topics in Quantum Electronics, IEEE Journal of*, PP(99):1, 2012a. ISSN 1077-260X. DOI: 10.1109/JSTQE.2012.2209113. 71

Roberto Proietti, Yawei Yin, Runxiang Yu, Xiaohui Ye, Christopher Nitta, Venkatesh Akella, and S. J. B. Yoo.   All-optical physical layer NACK in awgr-based optical interconnects. *Photonics Technology Letters, IEEE*, 24(5):410 –412, March 2012b. ISSN 1041-1135. DOI: 10.1109/LPT.2011.2179923. 71

Yusheng Qian, Seunghyun Kim, Jiguo Song, Gregory P. Nordin, and Jianhua Jiang. Compact and low loss silicon-on-insulator rib waveguide 90° bend. *Optics Express*, 14(13):6020–6028, June 2006. DOI: 10.1364/OE.14.006020. 26

Vivek Raghunathan, Winnie N. Ye, Juejun Hu, Tomoyuki Izuhara, Jurgen Michel, and Lionel Kimerling.   Athermal operation of silicon waveguides: spectral, second order and footprint dependencies.   *Optics Express*, 18(17):17631–17639, August 2010.   URL http://www.opticsexpress.org/abstract.cfm?URI=oe-18-17-17631. DOI: 10.1364/OE.18.017631. 52

Jonathan Schrauwen, Dries Van Thourhout, and Roel Baets. Trimming of silicon ring resonator by electron beam induced compaction and strain. *Optics Express*, 16(6):3738–3743, 2008. DOI: 10.1364/OE.16.003738. 51

Shankar Kumar Selvaraja. *Wafer-Scale Fabrication Technology for Silicon Photonic Integrated Circuits*. PhD thesis, Ghent University, 2011. 51, 52

Semiconductor Industry Association. International technology roadmap for semiconductors 2011. Technical report, Semiconductor Industry Association, 2011. URL http://www.itrs.net/Links/2011ITRS/Home2011.htm. 2, 27, 30

Semiconductor Industry Association. International technology roadmap for semiconductors 2012. Technical report, Semiconductor Industry Association, 2012. URL http://www.itrs.net/Links/2012ITRS/Home2012.htm. 67

Assaf Shacham and Keren Bergman. Building ultralow-latency interconnection networks using photonic integration. *IEEE Micro*, 27(4):6–20, 2007. ISSN 0272-1732. DOI: 10.1109/MM.2007.64. 36, 38, 48

Assaf Shacham, Keren Bergman, and Luca P. Carloni. The case for low-power photonic networks on chip. In *DAC '07: Proceedings of the 44th annual Design Automation Conference*, pages 132–135, New York, NY, USA, 2007a. ACM. ISBN 978-1-59593-627-1. DOI: 10.1145/1278480.1278513. 36, 38, 48

Assaf Shacham, Keren Bergman, and Luca P. Carloni. On the design of a photonic network-on-chip. In *NOCS '07: Proceedings of the First International Symposium on Networks-on-Chip*, pages 53–64, Washington, DC, USA, 2007b. IEEE Computer Society. ISBN 0-7695-2773-6. DOI: 10.1109/NOCS.2007.35. 36, 38, 48

Assaf Shacham, Keren Bergman, and Luca P. Carloni. Photonic networks-on-chip for future generations of chip multiprocessors. *Computers, IEEE Transactions on*, 57(9):1246–1260, 2008. ISSN 0018-9340. DOI: 10.1109/TC.2008.78. 48, 49, 50

Michael J. Shaw, Junpeng Guo, Gregory A. Vawter, Scott Habermehl, and Charles T. Sullivan. Fabrication techniques for low-loss silicon nitride waveguides. In *Society of Photo-Optical Instrumentation Engineers (SPIE) Conference Series*, volume 5720 of *Society of Photo-Optical Instrumentation Engineers (SPIE) Conference Series*, pages 109–118, 2005. URL http://dx.doi.org/10.1117/12.588828. DOI: 10.1117/12.588828. 63

Nicolás Sherwood-Droz, Kyle Preston, Jacob S. Levy, and Michal Lipson. Device guidelines for WDM interconnects using silicon microring resonators. In *WINDS 2010 Workshop on the Interaction between Nanophotonic Devices and Systems*, pages 15–17, December 2010. 11, 15

Guoyong Sun, Dae Seung Moon, Aoxiang Lin, Won-Taek Han, and Youngjoo Chung. Tunable multiwavelength fiber laser using a comb filter based on erbium-ytterbium co-doped polarization maintaining fiber loop mirror. *Optics Express*, 16(6):3652–3658, March 2008. URL http://www.opticsexpress.org/abstract.cfm?URI=oe-16-6-3652. 8

Dirk Taillaert, Peter Bienstman, and Roel Baets. Compact efficient broadband grating coupler for silicon-on-insulator waveguides. *Optics Letters*, 29(23):2749–2751, 2004. URL http://ol.osa.org/abstract.cfm?URI=ol-29-23-2749. DOI: 10.1364/OL.29.002749. 15

Kazumasa Takada, Masanori Abe, Masaru Shibata, Motohaya Ishii, and Katsunari Okamoto. Low-crosstalk 10-GHz-spaced 512-channel arrayed-waveguide grating multi/demultiplexer fabricated on a 4-in wafer. *Photonics Technology Letters, IEEE*, 13(11):1182–1184, 2001. ISSN 1041-1135. DOI: 10.1109/68.959357. 71

Michael Tan, Paul Rosenberg, Jong Souk Yeo, Moray McLaren, Sagi Mathai, Terry Morris, Joseph Straznicky, Norman P. Jouppi, Huei-Pei Kuo, Shih-Yuan Wang, Scott Lerner, Pavel Kornilovich, Neal Meyer, Robert Bicknell, Charles Otis, and Len Seals. A high-speed optical multi-drop bus for computer interconnections. In *High Performance Interconnects, 2008. HOTI '08. 16th IEEE Symposium on*, pages 3–10, 2008. DOI: 10.1109/MM.2009.57. 69

Liang Tang and David A. B. Miller. Commentary: Metallic nanodevices for chip-scale optical interconnects. *Journal of Nanophotonics*, 3(1):1–4, March 2009. DOI: 10.1117/1.3111849. 66

Jie Teng, Pieter Dumon, Wim Bogaerts, Hongbo Zhang, Xigao Jian, Xiuyou Han, Mingshan Zhao, Geert Morthier, and Roel Baets. Athermal silicon-on-insulator ring resonators by overlaying a polymer cladding on narrowed waveguides. *Optics Express*, 17(17):14627–14633, August 2009. URL http://www.opticsexpress.org/abstract.cfm?URI=oe-17-17-14627. DOI: 10.1364/OE.17.014627. 52

Hon Ki Tsang, C. S. Wong, T. K. Liang, Ian E. Day, Stephen W. Roberts, Arnold Harpin, John P. Drake, and Mehdi Asghari. Optical dispersion, two-photon absorption and self-phase modulation in silicon waveguides at 1.5μm wavelength. *Applied Physics Letters*, 80(3):416–418, 2002. ISSN 0003-6951. DOI: 10.1063/1.1435801. 15

Aniruddha N. Udipi, Naveen Muralimanohar, Rajeev Balasubramonian, Al Davis, and Norman P. Jouppi. Combining memory and a controller with photonics through 3D-stacking to enable scalable and energy-efficient systems. In *Proceedings of the 38th Annual International Symposium on Computer Architecture*, ISCA '11, pages 425–436, New York, NY, USA, 2011. ACM. ISBN 978-1-4503-0472-6. URL http://doi.acm.org/10.1145/2000064.2000115. DOI: 10.1145/2024723.2000115. 69

Amin Vahdat. Delivering scale out data center networking with optics; Why and how. In *Optical Fiber Communication Conference and Exposition (OFC/NFOEC), 2012 and the National Fiber Optic Engineers Conference*, pages 1–36, March 2012. DOI: 10.1364/OFC.2012.OTu1B.1. 2, 38

Amin Vahdat, Hong Liu, Xiaoxue Zhao, and Chris Johnson. The emerging optical data center. In *Optical Fiber Communication Conference and Exposition (OFC/NFOEC), 2011 and the National*

Fiber Optic Engineers Conference, pages 1–3, March 2011. DOI: 10.1364/OFC.2011.OTuH2. 2, 38

Joris Van Campenhout, Pedro Rojo Romeo, Philippe Regreny, Christian Seassal, Dries Van Thourhout, Steven Verstuyft, Lea Di Cioccio, Jean-Marc Fédéli, Chrystelle Lagahe, and Roel Baets. Electrically pumped InP-based microdisk lasers integrated with a nanophotonic silicon-on-insulator waveguide circuit. *Optics Express*, 15(11):6744–6749, May 2007. URL http://www.opticsexpress.org/abstract.cfm?URI=oe-15-11-6744. DOI: 10.1364/OE.15.006744. 64

Joris Van Campenhout, Pedro R. Romeo, Dries Van Thourhout, Christian Seassal, Philippe Regreny, Lea Di Cioccio, Jean-Marc Fedeli, and Roel Baets. Design and optimization of electrically injected InP-based microdisk lasers integrated on and coupled to a SOI waveguide circuit. *Lightwave Technology, Journal of*, 26(1):52–63, 2008. ISSN 0733-8724. DOI: 10.1109/JLT.2007.912107. 64

Dana Vantrease, Robert Schreiber, Matteo Monchiero, Moray McLaren, Norman P. Jouppi, Marco Fiorentino, Al Davis, Nathan Binkert, Raymond G. Beausoleil, and Jung Ho Ahn. Corona: System implications of emerging nanophotonic technology. In *ISCA '08: Proceedings of the 35th International Symposium on Computer Architecture*, pages 153–164, Washington, DC, USA, 2008. IEEE Computer Society. ISBN 978-0-7695-3174-8. DOI: 10.1109/ISCA.2008.35. 14, 21, 26, 27, 36, 38, 40, 41, 45, 56, 69

Dana Vantrease, Nathan Binkert, Robert Schreiber, and Mikko H. Lipasti. Light speed arbitration and flow control for nanophotonic interconnects. In *Micro-42: Proceedings of the 42nd Annual IEEE/ACM International Symposium on Microarchitecture*, pages 304–315, New York, NY, USA, 2009. ACM. ISBN 978-1-60558-798-1. DOI: 10.1145/1669112.1669152. 40, 41, 45

Dana M. Vantrease. *Optical Tokens in Many-Core Processors*. PhD thesis, University of Wisconsin-Madison, 2010. 16, 23, 41

Hassan M. G. Wassel, Mohit Tiwari, Jonathan K. Valamehr, Luke Theogarajan, Jennifer Dionne, Frederic T. Chong, and Timothy Sherwood. Towards chip-scale plasmonic interconnects. In *WINDS 2010 Workshop on the Interaction between Nanophotonic Devices and Systems*, pages 23–24, December 2010. 65

Hassan M. G. Wassel, Daoxin Dai, Mohit Tiwari, Jonathan K. Valamehr, Luke Theogarajan, Jennifer Dionne, Frederic T. Chong, and Timothy Sherwood. Opportunities and challenges of using plasmonic components in nanophotonic architectures. *Emerging and Selected Topics in Circuits and Systems, IEEE Journal on*, 2(2):154–168, 2012. ISSN 2156-3357. 65, 66

Gregory L. Wojcik, Dongliang Yin, Alexey R. Kovsh, Alexey E. Gubenko, Igor L. Krestnikov, Sergey S. Mikhrin, Daniil A. Livshits, David A. Fattal, Marco Fiorentino, and Raymond G. Beausoleil. A single comb laser source for short reach WDM interconnects. In *Proceedings of SPIE*, volume 7230, pages 72300M–72300M–12, 2009. URL http://dx.doi.org/10.1117/12.816278. DOI: 10.1117/12.816278. 64

Dong Hyuk Woo, Nak Hee Seong, Dean L. Lewis, and Hsien-Hsin S. Lee. An optimized 3D-stacked memory architecture by exploiting excessive, high-density TSV bandwidth. In *High Performance Computer Architecture (HPCA), 2010 IEEE 16th International Symposium on*, pages 1–12. IEEE, 2010. DOI: 10.1109/HPCA.2010.5416628. 4

Qianfan Xu, David Fattal, and Raymond G. Beausoleil. Silicon microring resonators with 1.5-μm radius. *Optics Express*, 16(6):4309–4315, 2008. DOI: 10.1364/OE.16.004309. 11

Xiaohui Ye, Yawei Yin, S. J. Ben Yoo, Paul Mejia, Roberto Proietti, and Venkatesh Akella. DOS - A scalable optical switch for datacenters. In *Architectures for Networking and Communications Systems (ANCS), 2010 ACM/IEEE Symposium on*, ANCS '10, pages 1–12, 2010. ISBN 978-1-4503-0379-8. DOI: 10.1145/1872007.1872037. 71

S. J. Ben Yoo. Optical packet and burst switching technologies for the future photonic internet. *Lightwave Technology, Journal of*, 24(12):4468–4492, 2006. ISSN 0733-8724. DOI: 10.1109/JLT.2006.886060. 71

Ian A. Young, Edris Mohammed, Jason T. S. Liao, Alexandra M. Kern, Samuel Palermo, Bruce A. Block, Miriam R. Reshotko, and Peter L. D. Chang. Optical I/O Technology for Tera-Scale Computing. *Solid-State Circuits, IEEE Journal of*, 45(1):235 –248, January 2010. ISSN 0018-9200. DOI: 10.1109/ISSCC.2009.4977511. 2, 67

Leila Yousefi and Amy C. Foster. Waveguide-fed optical hybrid plasmonic patch nano-antenna. *Optics Express*, 20(16):18326–18335, Jul 2012. URL http://www.opticsexpress.org/abstract.cfm?URI=oe-20-16-18326. DOI: 10.1364/OE.20.018326. 16, 66

Linjie Zhou, Ken Kashiwagi, Katsunari Okamoto, R. Scott, N. Fontaine, Dan Ding, Venkatesh Akella, and S. Yoo. Towards athermal optically-interconnected computing system using slotted silicon microring resonators and RF-photonic comb generation. *Applied Physics A: Materials Science & Processing*, 95:1101–1109, 2009a. ISSN 0947-8396. DOI: 10.1007/s00339-009-5120-7. 65

Linjie Zhou, Katsunari Okamoto, and S. J. Ben Yoo. Athermalizing and trimming of slotted silicon microring resonators with UV-sensitive PMMA upper-cladding. *Photonics Technology Letters, IEEE*, 21(17):1175–1177, September 2009b. ISSN 1041-1135. DOI: 10.1109/LPT.2009.2023522. 3, 51, 52

Authors' Biographies

CHRISTOPHER JOHN NITTA

Christopher Nitta received his Ph.D. in Computer Science from the University of California, Davis and is an adjunct professor of Computer Science at the University of California, Davis. His research interests include network-on-chip technologies, embedded system and RTOS design, and hybrid electric vehicle control.

MATTHEW KARL FARRENS

Matthew Farrens received his Ph.D. in Electrical Engineering from the University of Wisconsin and is a professor of Computer Science at the University of California, Davis. His research interests center on computer architecture, with special emphasis on the memory hierarchy. He is a member of ACM and IEEE and a recipient of the NSF PYI award.

VENKATESH AKELLA

Venkatesh Akella received his Ph.D. in Computer Science from University of Utah and is a professor of Electrical & Computer Engineering at University of California, Davis. His current research encompasses various aspects of embedded systems and computer architecture with special emphasis on embedded software, hardware/software codesign and low power system design. He is member of ACM and received the NSF CAREER award.